The Political Landscape

The Political Landscape

Constellations of Authority
in Early Complex Polities

Adam T. Smith

UNIVERSITY OF CALIFORNIA PRESS

Berkeley / Los Angeles / London

University of California Press
Berkeley and Los Angeles, California

University of California Press, Ltd.
London, England

© 2003 by
The Regents of the University of California

Library of Congress Cataloging-in-Publication Data

Smith, Adam T.
    The political landscape : constellations of authority in early
complex polities / Adam T. Smith.
        p.   cm.
Includes bibliographical references (p.   ) and index.
    ISBN 0-520-23749-8 (cloth : alk. paper)—ISBN 0-520-23750-1
(pbk. : alk. paper)
    1. Political anthropology.   2. Landscape archaeology—Political
aspects.   3. Landscape assessment.   I. Title.
GN492.2 .S64   2003
306.2—dc21   2003005058

Manufactured in the United States of America
12   11   10   09   08   07   06   05   04   03
10   9   8   7   6   5   4   3   2   1

To my parents, with love and thanks

Contents

Illustrations

Figures

Table

Acknowledgments

Perhaps because this book is concerned with the role that landscapes play in our lives, it is difficult for me to reflect on the many individuals who contributed to the manuscript except in terms of the places where the book took shape. The present study owes a profound debt to the institutions where it was pursued and the numerous friends, colleagues, and students who offered thoughtful critical reflections on the manuscript as it moved through various drafts. The work began as a theoretical problem that developed out of the dissertation research that I conducted in southern Transcaucasia while a student at the University of Arizona in Tucson. T. Patrick Culbert, David Killick, and Ana Alonso all played important roles in developing the conceptual perspectives that drove me to explore the implications of emerging spatial theory for an archaeology of politics. I owe a particularly sizable debt to the late Carol Kramer, who not only supervised my research but also took on the role of mentor. It is with great sadness that I contemplate completing this work without her. And I grieve for the future students of anthropology now deprived of her knowledge, wit, independence, and dedication.

The research project expanded dramatically while I was a post-doctoral fellow in the Society of Fellows at the University of Michigan, Ann Arbor. What had begun as an inquiry into Urartian uses of landscape developed into a project in comparative political anthropology that sought to examine overlapping sets of issues in Mesopotamia and Mesoamerica. A number of colleagues at Michigan played profound roles in shaping the analytical dimensions of this book. Susan Alcock, John Cherry, Piotr Michalowski, Jeffrey Parsons, and Henry Wright all provided a great deal of assistance in thinking through various elements of the project, from the highly detailed to the broadly theoretical. I particularly want to thank Norman Yoffee for serving as that most sought after of colleagues: the friendly, supportive, and careful critic. I owe him an incalculable debt for his guidance and support. I would also like to thank those students of Professor Yoffee who read and commented on early drafts of the present work. And I must convey my gratitude to the Michigan Society of Fellows, an extraordinary institution, for offering an exceedingly pleasant environment for scholarly work and discussion. A number of my colleagues offered very useful comments on portions of the text and the overall design of the work, including Paul Anderson, Alaina Lemon, and James Boyd White.

Only after I joined the faculty of the University of Chicago did the manuscript reach fruition. I am indebted to many colleagues and students who offered trenchant criticism on various drafts of chapters or provided critical rejoinders standing in the corridors of Haskell Hall. The participants in my graduate seminars on "Landscape" and "The Archaeology of Political Life" provided valuable insights into the shape of the theoretical argument. Helpful comments were also provided by participants in the Chicago Theoretical Archaeology Workshop. I want to thank Ian Straughn, who read large swaths of the manuscript at a very early stage; Patrick Scott, who proved a thoughtful critic in the Theoretical Archaeology Workshop; and Peter Johansen, who provided cogent commentary that helped to refine my thinking on several issues.

Many of my colleagues offered exceedingly useful advice and critiques of various arguments and points developed in the text, including Andrew Apter, Michael Dietler, Nadia Abu el-Haj, Alan Kolata, and Nicholas Kouchoukos. John L. Comaroff gave several chapters a close, critical reading that helped me to tighten my argumentation and fill in gaps in my understanding of the State. I particularly want to thank Robert McC. Adams, who provided an important boost of encouragement as the manuscript entered its final form as well as numerous insightful comments and thoughtful suggestions. In addition, several colleagues in the Oriental Institute have been extremely generous with their time and their ideas. Tony Wilkinson provided very thoughtful reflections and comments on the problems of landscape as viewed from the Near East. MacGuire Gibson, John Saunders, and John Larsen provided much appreciated help with key illustrations.

A word of acknowledgment must also be extended to numerous colleagues from diverse places who offered extensive comments and critiques that forced me to argue various issues much more closely. I owe a great debt first and foremost to Ruben Badalyan of the Institute of Archaeology and Ethnography in Yerevan, Armenia, who has been an insightful colleague and supportive friend since I first started conducting fieldwork in Caucasia. I also want to thank Pavel Avetisyan, also of the Institute of Archaeology and Ethnography in Yerevan, for his suggestions and thoughts on various issues raised in the following pages. My thanks to Philip Kohl for first introducing me to the archaeology of southern Caucasia. Both Cynthia Robin and Patricia Cook provided a great deal of expertise on Maya archaeology, without which I would undoubtedly have been lost. Geoff Emberling offered helpful reactions to my account of landscape, particularly as applied to early Mesopotamia. Ruth Van Dyke provided a great deal of extremely helpful commentary on issues of space and landscape. And I must also thank the two anonymous reviewers who read the manuscript for the University of California Press and provided me with highly detailed readings of the manuscript that helped both in my argumentation and in the construction of the narrative.

Throughout the process of writing this book, I have been guided by the staff of the University of California Press, including Marian Olivas, Suzanne Knott, Julia Johnson Zafferano, and Stanley Holwitz. Stan was remarkably patient as this book gained in intensity and argumentation even as it missed a couple of deadlines.

Finally, I must thank my family for their patience and enthusiasm. It can fairly be said that without my wife, Amy, this project would never have been completed. No acknowledgment can express my thanks for her patience, love, and support. Without my parents, who stoked my abiding interest in politics (albeit with substantially different ends in mind!), this project would never have begun. And it is with thanks for beginnings that I dedicate this book to them.

Introduction

Surveying the Political Landscape

Perhaps never in two thousand years has the reality of the state been so dim in men's minds.

Richard Wright, "Two Letters to Dorothy Norman"

In 1928, the *Illustrated London News* published a sensational pair of images based on C. Leonard Woolley's excavations at the ancient Mesopotamian city of Ur that seemed to capture political authority at the very instant of its reproduction.[1] The precociously cinematic illustrations depicted the tomb of Queen Puabi at a moment in the mid-third millennium B.C. when the retainers of the recently dead royal were assembled in the "Great Death Pit," preparing to accompany the queen into the afterlife.[2] In the first image, guards, servants, oxen, and carts are set in place around the vaulted chamber of the interred queen (fig. 1a). Although the individual figures in the scene appear rather stiff, the effect of the tableau is one of anticipation; the lack of movement presages a dramatic denouement. The grisly succeeding image portrays the climactic resolution of the scene (fig. 1b). Woolley's excavations revealed slaughtered animals

---

1. The *Illustrated London News* ran no fewer than 30 reports on Woolley's excavations at Ur over years of active excavations at the site (Zettler 1998: 9).

2. Several texts believed to bear directly upon third millennium B.C. Mesopotamian burial practices, such as *The Death of Gilgamesh* and *The Death of Ur-Namma*, indicate that deities and rulers could have palaces in the afterlife and that the burial of the retinue was to enable the departed to "continue living in the style to which he [or she] was accustomed" (Tinney 1998: 28).

FIGURE 1. The death pit of Queen Puabi of Ur as reconstructed by A. Forestier of the *Illustrated London News* (June 23, 1928). **a.** In the pit, attendants, oxen, and carts are arrayed in front of the vaulted tomb of the dead queen. **b.** (opposite) After the sacrifice, the arrayed bodies of humans and animals litter the pit while a cutaway of the tomb reveals the interred royal. (Courtesy of the *Illustrated London News* picture library.)

and poisoned attendants littering the floor of Puabi's antechamber, a spectacle of consumption that reinforced the power of the royal regime to command and the dedication of Ur's civil community to the existing political order. Although the violence of the scene gives the illustrations a voyeuristic quality, what is most intriguing about Puabi's tomb is not the brutality of political authority at work. Rather, Woolley's excavations in the Royal Cemetery at Ur revealed a vision of political authority that was firmly located in the persons and apparatus of the royal regime, a spatial immediacy conveyed in the illustrations by the backgrounded tomb. Politics in ancient Ur, the tomb of Puabi indicates, was set firmly in place.

A more chilling vision of the physical sacrifice of political subjects at the hands of ruling authority is found in Andy Warhol's (1965) silk screens of the electric chair at Sing Sing prison (fig. 2). Here we find a modernist vision of authority established by the absence both of the condemned (creating an ominous sense of the potential insertion of the viewer) and of authorities (present only as the mechanized apparatus of capital punishment).

The sign on the wall demanding silence accomplishes the final effacement of both the political subject (rendered mute even at the final moment) and the political regime itself, which vanishes behind the instruments of routinization. The series of silk screens works most powerfully in the dramatic repetition of the same image, washed with shifting color tones. In this repetition, Warhol created a powerful image of the modernist State's relation to its subjects—possessed of the same authority to command the ultimate sacrifice of political subjects as the kings and queens of Ur, yet profoundly unlocatable, simultaneously nowhere and everywhere.

FIGURE 2. Andy Warhol, *Electric Chair* (1965). Silkscreen ink on synthetic polymer paint on canvas, 22 × 28 in. (The Andy Warhol Museum, Pittsburgh; Founding Collection. Contribution of the Andy Warhol Foundation for the Visual Arts, Inc. Photo: Richard Stoner.)

The death pit of Queen Puabi and Warhol's electric chair present images of the reproduction of political authority through the sacrifice of what Michel Foucault (1979a: 138) called the "docile bodies" of subjects. In the former, authority is quite close at hand, locatable in the present body of the royal contained within the architecture of her tomb. In the latter, power is vested in a technology of execution, but the place of authority is entirely obscured. The State, as Richard Wright complained in 1948, has indeed dimmed in the political imagination during the epochs separating Puabi's death pit and Sing Sing's electric chair. Within modern conceptions of the political, it has become increasingly difficult to define where authority is located, to understand not only (meta)historical transformations in regimes—the resolute focus of modernist inquiry—but also the constitution of authority at the intersection of both space and time. Wright's suggestion that the Cheshire Cat–like disappearance of the State boasts a genealogy far deeper than the modern era opens analytical space for an archaeological approach to the problem, one that juxtaposes the contemporary concerns of political thought with interpretations of land-scapes forged by early complex polities. More than any other of the social sciences, archaeological perspectives on political life must directly confront the difficulties posed by understanding authority through places — in the ruins of built environments, distributions of artifacts, and images of town and country.

This book addresses the constitution of political authority in space and time through the making and remaking of landscapes in early complex polities—ancient political formations in which authority was predicated on radical social inequality, legitimated in reference to enduring representations of civil community, and vested in centralized organs of governance— in order to re-emplace the operation of political power and the manufacture of political legitimacy by governing regimes. (See chapter 2 for an extended definition of "early complex polity.") How do landscapes— defined in the broadest sense to incorporate the physical contours of the created environment, the aesthetics of built form, and the imaginative reflections of spatial representations—contribute to the constitution of political authority? Are the cities and villages in which we live and work, the lands that are woven into our senses of cultural and personal identity, and the national territories within which we are subjects merely stages on which historical processes and political rituals are enacted? Or do the forms of buildings and streets, the evocative sensibilities of architecture and vista, and the discursive aesthetics of place conjured in art and media consti-tute *political landscapes*—broad sets of spatial practices critical to the for-mation, operation, and overthrow of geopolitical orders, of polities, of regimes, of institutions? The central concern behind these questions is not the description and elaboration of landscapes in and of themselves, but how politics operates through landscapes.

A Topology of Political Landscapes: Defining Terms

In opening a discussion of landscapes in political life, we might reflect on the following scenes:

MOSCOW, October 4, 1993 *(New York Times)*—Dozens of tanks and armored personnel carriers loyal to President Boris Yeltsin today bombarded the rebel-held Russian parliament building, after thousands of armed anti-government demonstrators had routed police Sunday, seized key buildings in the capital, and fought a pitched battle with guards at the state television complex.

BELFAST, July 19, 1998 (CNN.com) — Police and soldiers, enforcing an order
by the government-appointed Parades Commission, prevented members of
the Orange Order, Northern Ireland's largest Protestant brotherhood, from
marching down the Garvaghy Road in Portadown, 25 miles (40 km) southwest of Belfast, on July 5. Orangemen massed on the hillside near a rural church
every night for a week, underlining their determination to follow their traditional parade route.

JERUSALEM, November 16, 1998 (CNN.com) — In comments broadcast on
Israeli radio, [then defense minister Ariel] Sharon urged Jewish settlers to grab
West Bank hilltops before a permanent agreement is reached on the area where
Palestinians hope to build an independent homeland. "Everyone there should
move, should run, should grab more hills, expand the territory," exhorted
Sharon. "Everything that's grabbed will be in our hands. Everything we don't
grab will be in their hands."

In each of these episodes, the created environment is not simply a backdrop for political activities but rather the very stake of political struggle. It
is the spaces themselves—the Russian parliament building, Garvaghy Road,
the hilltops of the West Bank—and the real and imagined landscapes in
which they are situated—a post-Soviet Russia, an autonomous Northern
Ireland, an independent Palestine—that are fought over, argued over, and
died for. During such moments of intense crisis, the spatiality of political
authority comes into focus with striking clarity as the security of regimes,
the integrity of polities, and the reproduction of national communities hinge
on the resolution of problems of spatial form, extent, distribution, and representation.[3] Similarly, the contestation of political authority—resistance
to its commands and attempts to seize its controls—relies on a cartographic
vision, a spatialized understanding of the sites of singular importance to
reproduction and transformation. It is this spatialized understanding of
movements against the dominant political apparatus that Georges Bataille
struck on when he suggested that the storming of the Bastille in 1789 must
be understood as an expression of "l'animosité du peuple contre les monuments qui sont ses véritables maîtres [the animosity of the people against
the monuments that are their true rulers]" (1929: 117).

The prominence of the landscape in episodes of authority in crisis begs
the more general question of the everyday significance of space and place
in civil life. What currency can the concept of a political landscape have
in the quotidian realm of everyday politics? The phrase "political land-

3. Considerable anthropological attention has recently been given to the spatiality of
crises in Northern Ireland (e.g., Feldman 1991).

scape" is deceptively multivalent for a term as familiar to pundits as scholars. As a colloquial expression, the phrase generally refers to the array of governmental and opposition forces at a given point in time. This is the sense of the phrase used in political reportage, generating a host of articles describing the political landscapes of Europe, of the U.S. health-care debate, of the Republican Party, and so forth (see, for example, Gottfried 1995; Burka 1994). It is to this purely metaphoric denotation that Michelle Mitchell (1998) appeals in her study of the impact of Generation X on traditional political allegiances, subtitled *How the Young Are Tearing Up the American Political Landscape.* Although useful in journalistic contexts, the analytical utility of this sense of political landscape is limited by the shallowness of the spatial metaphor. That is, landscape in this sense refers only to a stylized sketch of intersecting political forces, as useful in its tropic conventions as "political climate." However, the impetus to create even a caricatured cartography descriptive of allied and opposed political forces suggests that the phrase is worth unpacking.

There is a more literal sense of the political landscape, one that describes the physical ordering of the created environment by political forces. It is this sense of the term that J. B. Jackson (1970) employed in a lecture delivered at the University of Massachusetts in 1966. For Jackson, founder of the journal *Landscape,* the political landscape describes the "megastructure" of the American created environment, a skeleton composed of highways, boundaries, meeting places, and monuments on which all other segments of the spatial order are hung. Jackson's vision of the political landscape is teleological in that its features emerge in relation to its ends rather than in the process of its formation: "the political landscape, the landscape designed to produce law-abiding citizens, honest officials, eloquent orators, and patriotic soldiers" (1984: 39). One practical criticism that can be leveled at Jackson's vision is that his sense of what physical forms might be traced to political sources was far too restricted. As studies such as Mike Davis's *City of Quartz* (1990) and James Kunstler's *The Geography of Nowhere* (1993) make clear, political action can be located throughout the created environment: in the form of a park bench designed to prevent homeless people from intruding on gentrifying neighborhoods, in the zoning rules that push homes back from the street and encourage the sprawl of strip malls, in the gridded layout of streets and access routes. Indeed, in a pedestrian sense we might consider how each time a red light halts our progress, we are interpellated (in an Althusserian [1971] sense) as subjects of a mechanized authority codified by the instrumentality of the political landscape.

To press beyond the purely experiential sense of the term, a denotation centered on the flow of goods and bodies, "political landscape" also carries a more semiotic sensibility, one that employs politically generated signs to shape our sense of place. For example, the most prominently iterated value shaping the establishment of public spaces in the United States during the late nineteenth century was the attempt to create democratic places where middle-class values might be paraded for the edification and encouragement of lower social strata (Olmsted 1971). In contrast, the central value orienting the configuration of public space today is the insulation of that same middle class from contact with the poor. A wide range of recent policies has been designed to reduce the openness of public gathering places by physically removing or intimidating the undesirable segment of the population (such as former New York City mayor Rudolph Giuliani's criminalization of homelessness through sweeps of popular New York tourist destinations) or by encouraging the public to congregate in privately held (but publicly subsidized) places that sequester the life of the city into massive office and shopping centers (such as Detroit's Renaissance Center, Atlanta's Peachtree and Omni Centers, and Los Angeles's Bunker Hill and the Figueroa corridor [Davis 1990]). This transformation in American attitudes toward public spaces has only been possible insofar as political authorities have come to embrace corporatism as a strategy for dealing with social problems.[4]

The embedding of civic values within the concept of landscape leads to a third sense of "political landscape." In this construction of the term, built features evoke affective responses, enlisting emotions generated by sensory responses to form and aesthetics in service to the polity. It is precisely this use of the term that James Mayo employs in his examination of American war memorials. As Mayo notes in framing his study, "War is the ultimate political conflict, and attempts to commemorate it unavoidably create a distinct political landscape. . . . As an artifact a memorial helps create an ongoing order and meaning beyond the fleeting and chaotic experiences of life" (1988: 1). In this use of the phrase, the political landscape is constituted in the places that draw together the imagined civil community, a perceptual dimension of space in which built forms elicit affective responses that galvanize memories and emotions central to the experience of political belonging. Here we might think of the triumph of the Vietnam War Memorial, which succeeds precisely because it eulogizes a

---

4. See Low 2000 for a discussion of the adverse impact of the erosion of public space upon democratic politics and society.

divisive episode in U.S. history in terms that elicit not a sharp rebuke of a failed policy but a rededication of observers to the civil community.[5]

A final reading of "political landscape" arises from the etymological roots of "landscape" to describe a genre of pictorial image and centers on the impact of representations on our imagination of political life. Described during the Renaissance as "of all kinds of painting the most innocent, and which the Divill himselfe could never accuse of idolatry" (Edwin Norgate, *Miniatura,* quoted in Gombrich 1966: 107; see also Mitchell 2000: 193), the genre of landscape painting was castigated in modernist circles as naïve in its ostensible ambition to provide a pure aesthetic rendering of a wholly natural world. Martin Warnke, Anne Bermingham, and W. J. T. Mitchell (among others) have led a recent reevaluation of the sources of landscape painting, arguing that such images assimilate a wholly created environment to the political ends of rulers: "The ruler may be portrayed as controlling the forces of nature; alternatively he may appear as the distributor of her gifts and so confirm her superior status" (Warnke 1995: 145). The political landscape in this sense describes a representation of space whose ordering aesthetic derives from the goals and ambitions of regimes. Warnke also describes how features of the built environment impinge on how we perceive those worlds. In describing one of the most prominent elements of European political landscapes, Warnke points out that "[c]astles not only use topographical features for practical purposes, but call for mental attitudes" (ibid.).

As Warnke's analysis suggests, we should not be surprised to find that not only do the physical features of the political landscape encourage affective responses in the viewers but they may also have a profound impact on how we imagine idealized landscapes. From the eerily panoptic medieval castle of the Disney theme parks that surveys a 1950s fantasy-scape Main Street USA to the purely digital geographies of Multi-User Domains (MUDs), our imagined landscapes regularly pivot around a central apparatus of political authority—a civil *axis mundi.*[6] The impact of these imagined spatial sensibilities, centered on the imminent presence of a po-

5. See Hass 1998 on the conflict that surrounded the construction of the Vietnam War Memorial and the role of national memory in understanding its continued efficacy.

6. A large number of digital inquiries into the consequences of cyberspace for traditional notions of city and territory can be found on the internet, including Hypernation (http://duplox.wz-berlin.de/netze/netzforum/archive.20may96/0201.html), Refugee Republic (www.refugee.net/), and the Amsterdam Digital City (www.dds.nl/dds/info/english/engelsfolder95.html). On the architecture of the Disney theme parks, see the papers in Marling 1997.

litical apparatus, is not hard to tease out from an American popular culture obsessed with conspiracies, eavesdropping, and the near omniscience credited to governments on television *(The X-Files)*, on film *(3 Days of the Condor, The Conversation, Enemy of the State, JFK, The Parallax View)*, and in literature *(Vineland, American Tabloid)*.[7]

To summarize, there are at least three senses of the term political landscape that provide a conceptual platform for the examination of the spatial constitution of civil authority:

· an imaginative aesthetic guiding representation of the world at hand;

· a sensibility evoking responses in subjects through perceptual dimensions of physical space; and

· an experience of form that shapes how we move through created environments.

These three dimensions of the political landscape organize the spatial studies of authority in early complex polities undertaken in the present work. Each of the studies explored in chapters 3 through 6 has been organized in tripartite fashion around experience, perception, and imagination in order to provide an encompassing description of political landscapes.

It is unsurprising that the phrase "political landscape" should be so multivalent given the plural meanings of its component terms. "Landscape" is difficult to define in part because it is so familiar. Used by artists, architects, developers, geologists, geographers, historians, and archaeologists (to name only a few), the term has varying denotations and connotations. Jackson (1979: 153) once sighed that, though he had spent a career talking and writing about landscapes, the concept continued to elude him. Denis Cosgrove (1993: 8–9) has cogently suggested that the term landscape refers to the totality of the external world as mediated through subjective human experience. In more direct terms, landscape is land transformed by human activity or perception. If land is an objective concept, a physical solid that composes the surface of the planet, then landscape can be understood as land that humans have modified, built on, traversed, or simply gazed upon. Because of this sense of human production that inheres in the term, landscape must be understood not simply as space or place but as a synthesis of spatiality and temporality. That is, because land-

7. It is important to note that this cartography of governmental panopticism is a postwar vision that contrasts markedly with the images of governmental loci as places of security that dominate much New Deal era literature and film (Szalay 2000).

scapes are "made," to use the term that W. G. Hoskins employed (so ~, borrowed by E. P. Thompson), they enclose both a diachronic sense of extent and a synchronic understanding of duration (see Hoskins 1977; Thompson 1963). The term thus resists the Hegelian trap of segregating time and space as analytical dimensions that, as we shall see, has bedeviled modernist accounts of political life.

We can thus disentangle the oft-conflated concepts of space, place, and landscape in the following terms. If space refers to the general concepts of extension and dimension that constitute form, then place, following the geographer Yi-Fu Tuan (1977), refers to how specific locales become incorporated into larger worlds of human action and meaning. Landscape then refers to the broad canvas of space and place constituted within histories of social and cultural life. Landscape arises in the historically rooted production of ties that bind together spaces (as forms delimiting physical experience), places (as geographic or built aesthetics that attach meanings to locations), and representations (as imagined cartographies of possible worlds).

A far more difficult term to pin down is the modifying adjective in our titular phrase, "political." Politics, in some form, encompasses or intrudes on every dimension of our lives. It lies not only with the administrators of institutions that constitute the most recognizable forms of governmental authority but also in the caseworkers and teachers that intercede in relations between parents and children or the regulations and procedures that delimit the links between employer and employee. And yet if the analysis of landscapes is to have any specificity to its objects, there is compelling reason to attempt—for the purposes of analysis—to delimit a sphere of the political within which we might trace spatial features back to specific sets of social relationships rooted in the production and reproduction of specific horizons of power and legitimacy.

There are a number of ways in which we might carve out a distinct political sphere to provide a specific locus for the present discussions. One approach would be to simply restrict the political to the apparatus of government. Although such an effort has difficulties even within the comparatively straightforward context of modern secular liberal democracies, it is largely unworkable within studies of early complex polities, where lines separating political, religious, military, economic, social, familial, and other forms of authority tend to be rather faintly drawn. Rather than aspire to set the political within a neatly carved out sociological sphere of formal locations, I will describe it within a specific set of relationships central to the production, maintenance, and overthrow of sovereign authority. These include geopolitical relationships among polities, ties between subjects and regimes that inscribe the polity, links among elites and "grassroots" organizations that constitute political regimes, and relations and rivalries among governing institutions.[8] The studies undertaken in chapters 3 through 6 engage with each of these sets of relationships in turn. I do not suggest that these four sets of relationships exhaust the political, merely that they are central to politics and civil life.[9]

The Topography of Political Landscapes: Philosophical Contexts

Despite the readily apparent spatiality of political life—the ways in which political life is created, fostered, challenged, broken down, and reconstituted in the production of fields, roads, buildings, parks, cities, communities, and polities—landscape has remained curiously underdeveloped in traditional accounts of the forms and transformations of authority. Keith Basso (1996: 31) has eloquently described how Western Apache histories attend much more profoundly to *where* events occurred rather than *when*, because place and placemaking root the past in the surrounding landscape. The opposite may be said for the dominant trends in social science, where approaches to political life have been primarily concerned with when (both in specific and in meta senses) and far less interested in where. Thus, it is important at the outset of this investigation to describe briefly what is at stake in developing a landscape perspective for contemporary theorizations of politics.

Dominant modernist accounts of political life explicitly marginalize space and place in explanation and critique. Foucault noted the general animosity of twentieth-century Marxism to the development of a spatial sensibility, recounting a public incident during which he was upbraided for his interest in the spatial dimensions of social life. His critic argued that "*space* is reactionary and capitalist, but *history* and *becoming* are revolutionary" (1984: 252, emphasis in original). This oft-cited encounter perfectly captures the central ontological premise suffused throughout the primary traditions of modern political thought. From Marxism and lib-

8. By "grassroots" I mean sociocultural organizations, such as kin groups, neighborhood groups, and professional societies, that often play critical supporting roles in providing vertical links between the elites of regimes and basal-level interest groups (see Mollenkopf 1992).
9. I use the terms civic and civil broadly to refer to the political community of a polity.

eral political theory to political sociology and anthropology, explanation and critique typically hinge on describing the transformation of polities over time. As the geographer Edward Soja has observed, "So unbudge-ably hegemonic has been this historicism of [the modern] theoretical con-sciousness that it has tended to occlude a comparable critical sensibility to the spatiality of social life" (1988: 10–11). Henri Lefebvre tracks the birth of this temporal privilege to Hegel, who codified the proposition that "his-torical time gives birth to that space which the state occupies and rules over" (Lefebvre 1991: 21). As the temporality of history assumes the cen-tral explanatory role in an understanding of political production, the spa-tiality of the State is demoted to a purely mechanical problem of engi-neering, with little enduring relevance to a general understanding of either the emergence of political forms or the imagining of proper civil orders. Although a thorough investigation of the relentless temporocentrism of modernist political theory is beyond the scope of this book, a brief overview should bring the problem into focus. (For expanded discussions of space and time in political thought, see Agnew 1999; Alonso 1994; Har-vey 1996; Howard 1998; Kuper 1972; Soja 1985, 1988.)

The temporality of political life was transformed into revolutionary history by Marx such that the critical spatial problems of material pro-duction were transposed into primarily historical questions:

> [W]e must begin by stating the first premise of all human existence and, there-fore, of all history . . . that men must be in a position to live in order to be able to "make history." But life involves before everything else eating and drinking, housing, clothing and various other things. The first *historical* act is thus the production of the means to satisfy these needs, the production of material life itself. And indeed this is an historical act, a fundamental condi-tion of history, which today, as thousands of years ago, must daily and hourly be fulfilled merely in order to sustain human life. (Marx 1986: 174, emphasis mine).

By recasting the complex spatiotemporal problems of procuring food and water and enclosing living areas as exclusively historical acts, Marx re-phrases the temporal privilege defined by Hegel in strictly material terms but does not alter the highly sublimated nature of its spatial presupposi-tions. As Marxism came to be elaborated in the twentieth century, space, as a dimension of social life, came to be not simply of marginal impor-tance but in some cases explicitly counter-revolutionary (as in Georg Lukács's [1971] suggestion that spatial difference serves to define the ex-tant false consciousness whose primacy falls away when a fully realized class consciousness provides a sublime view on the logic of history [see also Lefebvre 1991: 22]).

Although different in its emplotment, the historicism underlying modern liberal political thought provides a similarly underdeveloped account of the spatial dimensions of civil life. In the absence of a spatial account of politics, analysis and critique tend to become unmoored from the world at hand, forced to seek grounding in purely theoretical ideal situations. The result is a vision of politics divorced from any place, such as that advanced by John Rawls's canonical work, *A Theory of Justice*. Rawls describes the creation of the just society out of the choices made by fully rational agents whose national, ethnic, class, and gender identities are hidden behind a veil of ignorance. In other words, everything that might lend a sense of place is described a priori as an obstacle to justice. As John Gray (1992) noted in a fiercely polemical critique, Rawls's approach to political theory (which has largely set the agenda for the past 30 years [Ankersmit 1996: 3]) hinges on the misguided hope that all humans will shed allegiances and local identities to unite in a universal vision of the just political order— that space will finally recede as relevant to civic communities. Although Gray's indictment of Rawls grows organically from his post-Thatcherite resistance to the obliterating effects of the capitalist market, his contention that political thought must come to terms with spatial difference or else find itself unprepared for the current era dominated by reemergent national, religious, and ethnic particularisms certainly rings true.

For political sociology and anthropology, approaches to public life that attempt to provide a general understanding of the nature of civil society, the failure to articulate an account of how authority operates in landscapes has resulted in the reification of a set of analytical types and models as if they were real facets of political activity. A growing number of important studies have detailed how institutions, roles, and ruling groups come to be vested with authority. But in general, sociological and anthropological approaches to politics continue to muster such studies to amplify models of the evolution of the State over time rather than to explicate the workings of polities through landscapes as spaces produced, reproduced, and razed over time. (The most recent influential works in this long tradition include Harris 1979; Johnson and Earle 1987; Mann 1986; Sanderson 1990, 1995; Service 1975; Trigger 1998.) Political anthropology's sense of space has been traditionally drawn from mid-twentieth-century political geography, predicating its analytical units on segmented territories and neutralized locations that provide stages for performances and rituals (Fortes and Evans-Pritchard 1940; see also MacIver 1926).

Nancy Munn has documented both the disengagement of time from space in early-twentieth-century anthropology (citing Evans-Pritchard's privileging of an abstract "oecological time" in opposition to a parallel "oecological space" as a foundational intellectual position) and recent attempts to regain perspective on the co-constitution of time and space by putting "subjects back in their bodies" (Munn 1992: 97, 105). The rise of practice-oriented perspectives on actors and actions has brought with it a reconsideration of the links between space and time, most notably in the writings of Giddens and Bourdieu (see, for example, Bourdieu 1977; Giddens 1984). However, as Munn points out, neither structuration nor practice theory are particularly convincing in their claims to spatial sensitivity. Whereas the former fails to provide any real account of how space is implicated in either (re)producing structures or constraining agents, the latter reiterates the traditional privileging of time over space in strategic relationships among actors: the very context in which we might gain a new understanding of the spatiotemporal co-constitution of political life (Munn 1992: 106–8; Saunders 1989; Urry 1991: 160–61). Thus, despite a general movement within social and anthropological theory to focus on practices, actors, and actions, a formal reconsideration of the theory and analytics that have long separated space from time within the dominant traditions of social thought has not generally followed.[10] In the absence of such a transformation in the traditional intellectual parameters of political analysis in the social sciences, the social evolutionary perspective on the deeper history of modern states contends with few rivals.

However, three major consequences (at least) result from the enduring focus on the evolution of the State as the guiding backstory of contemporary political thought. First, as Philip Abrams (1988: 81–82) has argued, the State comes to be reified as a thing in itself, allowing the concept to act as a mask for the various messages and practices of domination and subjection that are at the heart of political production. Thus the State, as an analytical concept, is at best an illusory focus for research that lends coherence and continuity to a disparate set of authority relationships that are highly discontinuous; at worst, it is an instrument of domination in itself. Second, exclusive attention to the creation of the State in history leaves humans so divorced from everyday political production and reproduction that the stakes of transformation are far exceeded by the scale on which such changes can be generated. Only full-scale revolution—or,

10. Munn's (1986) remarkable study *The Fame of Gawa* points clearly toward a highly nuanced understanding of the intersection of space and time within sets of social practices.

even more debilitating from the perspective of political activism, a major transformation in the adaptive systems of surrounding environments— can bring about movement from one formal type of political organization to another. Such "superstition," as Roberto Unger (1997: 6) has observed, "encourages surrender" rather than a constructive engagement of political theory and practice. From such theoretical sources, it should be unsurprising that we find ourselves at the dawn of the twenty-first century without a plausible critical project for civil society. Third, by defining the State a priori as a product of evolution—of primarily metahistorical processes—the landscapes that manufacture and preserve the authority of regimes, that encourage the identification of a regime with the land, that organize the way alternative forms can be imagined, are dismissed as incidental to politics. As a result, political action is left without a sense of place that might provide a locus for orienting reflection on the contemporary order and for developing new possibilities for civil life.

The foregoing discussion is not intended to suggest that the dominant contemporary approaches to the political sphere have no sense of space. They most certainly do. But the spatial within these strains of modernist thought remains highly sublimated as a component of both theory and practice, occluded behind the juggernaut of temporal determination. The theoretical agenda of the present work is to illuminate the implicit descriptions of space that support the explanatory privileges accorded temporality in political thought and to shift the analytical terrain to recognize landscape as a conceptual locus for detailing the intersection of time and space. The temporality of landscape, I argue, is rooted not in meta-level transformations but in the highly practical procedures of production, reproduction, and reformation defined in interwoven sets of political relationships.

To leave the political unmoored from the landscape, to allow it to float across society and culture as a conceptual ghost ship simultaneously anywhere and nowhere, is to obscure the practical relations of authority that constitute the civil sphere. Without an account of the constitution of authority in the production of landscapes, political analysis drifts farther from everyday life, trading agency for determinism and imposed routines for general laws. By refusing to cede the landscape any role in processes of political formation, administration, and collapse, perceived regularities in structure are unduly amplified. Fifteenth century A.D. Venice, fifth century A.D. Tikal, and third millennium B.C. Uruk disappear into the singular category labeled "States" despite what would seem the rather salient facts of their highly variable form, geography, and spatial aesthetics. This is not to argue for any form of geographic or ecological determinism. But by bracketing even the most pedestrian facts of geographic and architectural difference, similarities in structure are imagined as regularities that inhere in a type; laws of political transformation gain priority over any account of how civil relationships actually worked.

Over the past 40 years, a great deal of writing and thought in various fields within the social sciences has concentrated on how to create a critical approach to the spatiality of politics. In attempting to break down the essential Hegelian historicism of thought about the State, writers such as Lefebvre, Soja, and David Harvey (about whom much more will be said in chapter 1) have assailed the epistemological premises on which evolutionary narratives had been predicated. Why then does the development of a critical sense of the production of authority in both space and time remain so thoroughly marginal to political thought? There are undoubtedly a number of answers to this question, but, even as the Hegelian temporal privilege is battered on epistemological and theoretical levels, perhaps the most formidable bulwark to the sublimated spatial consciousness remains—what Marx, and later Antonio Gramsci, called "real history" (Gramsci 1971: 182; Marx and Engels 1998: 43).

An Archaeology of Political Landscapes: (Meta)histories of the State

It is the simple glance backward from modern nation-states to the ancient world that provides the strongest foundation for despatialized approaches to contemporary politics. Both traditional historicism and social evolutionism, metahistorical programs originally outlined in the nineteenth century, provide a profound backstory for modern political formations in which the State—conceived of as both a coherent social type and a broadly defined apparatus of government—rises inexorably, if not inevitably, out of roots put down in prehistory. The most prominent story of the State, that provided by social evolutionism, unfolds over the course of six millennia in a handful of disparate regions from Mesopotamia to Mesoamerica. When transposed to diverse locales, its simplistic melody remains extraordinarily consistent.

The temporocentric story of politics begins with the Pristine State— an original, autochthonous political formation built on radical social inequality and centralized governmental institutions that emerged first on the alluvial plain between the lower Tigris and Euphrates Rivers and in

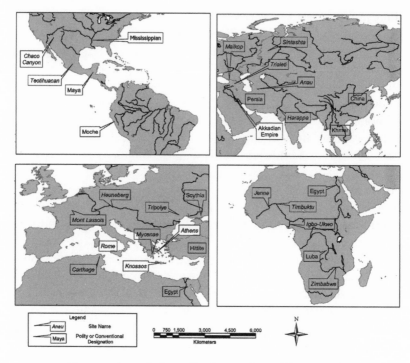

FIGURE 3. Four continental views of early complex polities: prominent sociopolitical horizons and key sites. (Map source: ESRI Data & Map CD.)

the Nile River valley. Sometime later, the Pristine State developed in a handful of other regions, including northern China, the Indus valley, Mesoamerica, and the Andes mountains (fig. 3). The Pristine State generally assumes one of two possible forms: regional state (for example, Old Kingdom Egypt) or city-state (for example, the interlinked urban polities of Early Dynastic southern Mesopotamia). As the Pristine State grew in complexity, it influenced surrounding regions either through imperial expansion or inter-regional political economy, thus sparking subsequent episodes of "secondary" State formation.[11] At the end of this ever-expanding network of secondary States lies the modern incarnation that,

11. For a recent effort to theorize the relations between fourth millennium B.C. southern Mesopotamian pristine states and polities on their periphery, see Algaze 2001a. For counter arguments that effectively question the explanatory integrity of traditional pristine/secondary divisions from the perspective of Mesopotamia's "periphery," see Frangipane 2001 and Stein 1999, 2001.

through its articulation of capitalist production and global colonialism, brings the State to its current position as *the* extant political formation.

The problem with this story does not rest in its archaeological facts; it does seem quite clear at present that the polities of southern Mesopotamia, Egypt, the Indus valley, northern China, Mesoamerica, and the Andes do represent the earliest experiments with complex orders of political authority in each region. (Depending on how we define both a geographic region and complexity, the same might be said of a large number of polities traditionally regarded as secondary States, including Scythia, Angkor, and Minoan Crete.) Rather, the weakness of the social evolutionary story of global State formation lies at its intellectual foundations, in its assumption that sufficient explanation of sociopolitical transformation resides at the intersection of general social types with metahistorical process. The spatial dimensions of political life, whether understood as the created environment in which politics takes place or the imagined spatial sensibilities behind the promulgation of new orders, are dismissed as epiphenomenal to the fundamental temporal axis of political transformation.[12]

A society's position in an evolutionary matrix, whether stringently or broadly conceived, is thought to determine its spatial configuration. Spatial forms are thus considered useful as benchmarks for social development but hold no interpretative status in themselves. This aspatial vision of political history comes in hard-line and more moderate forms, reaching its Whiggish apotheosis in a recent claim that, whereas restarting natural evolution would result in a whole new array of biological forms, a reprise of social evolution would engender the exact same results (Sanderson 1995: 7). Such breathtaking historical determinism follows inexorably from the over-privileging of temporality and reckless inattention to space that lies at the core of modernist political thought. Yet even if we set aside such examples of cold-blooded determinism, the narrative of the evolution of the State created out of the archaeological and historical records bolsters the modern disinterest in spatial dimensions of authority and thus inhibits any attempt to make the political sources of space and place relevant to discussions of civil life past, present, or future.

However, if we take a more archaeological perspective, the spatiality of political action can be located deep in human political history. One of

12. Contrast, for example, Aristotle's (1988) description of the spatial orders that obtain for each idealized constitutional form.

the earliest literary accounts of the origins of complex political communities opens with a recitation of the construction activities of King Gilgamesh at the city of Uruk in southern Mesopotamia—the earliest urban center in the world—hinting that place-making was considered to be *the* primordial political act.

> [Gilgamesh] built the rampart of Uruk-the-sheepfold,
> of holy Eanna, the sacred storehouse.
> See its wall like a strand of wool,
> view its parapet that none could copy!
> Take the stairway of a bygone era,
> draw near to Eanna, seat of Ishtar the goddess,
> that no later king could ever copy.
> Climb Uruk's wall and walk back and forth!
> Survey its foundations, examine the brickwork!
> Were its bricks not fired in an oven?
> Did the Seven Sages not lay its foundations?
>
> (George 1999: I.i)

The Epic of Gilgamesh (discussed at greater length in chapter 5) is separated from the episodes of political conflict that opened this introduction by at least 4,000 years, not to mention by the gossamer veil dividing myth from history. Yet they share a single intuition regarding the nature of political life: that the creation and preservation of political authority is a profoundly spatial problem. Gilgamesh's construction of the walls of Uruk is rendered as a technological marvel (as an architectural work), a pious devotion (to the glory of Ishtar), a genealogical fulfillment (of the work of the Seven Sages), and a triumph of settlement planning (through the quadripartite division of the city).

What emerges from the opening lines of the Gilgamesh epic is a sense that the space of Uruk was no less a politically constituted place than that of Moscow, the West Bank, or Garvaghy Road. Why then must places such as Mohenjo-Daro and Nineveh, Tikal, and Teotihuacan languish as poorly differentiated stages on which the State arises, reforms, and collapses? How can the spaces of modern Moscow be instrumental in preserving the authority of Yeltsin and Putin yet those of Babylon be superfluous to the authority of Hammurabi?

The investigations in this book are intended to demonstrate that political transformations in early complex polities were predicated on the production of very specific landscapes, thus undermining the "real" history that undergirds modernist temporocentric accounts of political life. Although I focus on ancient polities, the underlying concern of the work is the modern political formation in which both author and reader live. Indeed, the central conceit of the book is that political formations now visible primarily in ruins might provide the foundations for a modern critical project. This archaeological vision of politics warrants a brief discussion at the outset.

The Ancient and the Modern: Archaeology and Contemporary Politics

An overarching concern of *The Political Landscape* is to re-inject an archaeological perspective on political formation into what has largely become, to its detriment, a resolutely presentist discourse. It is not coincidental that, as the modern State has become increasingly dim in our minds, in Richard Wright's words, early complex polities have become increasingly distant and irrelevant. In order to combat this invisibility, it is vital that we attempt to bring the dimensions of early political life into better focus and thus provide a prism through which we might view contemporary civil authorities. But the intellectual link between archaeologies of complex polities and critiques of contemporary politics are not immediate or entirely straightforward. There are a number of lines along which we might develop a cogent argument for the utility of archaeological perspectives to contemporary political analysis and criticism. The first centers on the role of material culture within archaeological accounts of politics. In contrast to the ephemeral nature of the State generated within self-consciously postmodern social theory, archaeology's vision of politics has remained steadfastly centered on the intense physicality of power and governance. Although at times this perspective regretfully descends into naked materialism, this need not be the case, because the material culture of politics is as recursively instrumental in shaping the imagination as it is in regulating subsistence economies.

Within a more historical vein, the utility of an archaeological perspective can be framed in reference to the perhaps apocryphal quip of a late-twentieth-century Chinese diplomat. When asked to evaluate the significance of the French Revolution, he responded that it was as yet too soon to tell. The deep historical vision of archaeology has long been central to both its intellectual mission and its popular appeal. However, studies of politics—and political anthropology more specifically—have traditionally had a rather ambiguous relationship to archaeology. Social evolutionists, from Marx, Engels, and Lewis Henry Morgan to Elman Service, Marvin Harris, and Morton Fried, have long engaged with archaeological and epigraphic evidence to flesh out their idealized sequences; however, as evolutionism has fallen into disfavor, the historical vision within political anthropology has become increasingly myopic, focusing primarily on the rise of colonialism and modern capitalism. Intellectual ties between studies of early complex societies and anthropologies of modern life were severed under a presumed opposition between the historical particularities of global capitalism (the stuff of sociocultural anthropology) and the general metahistory of political evolution (the ambitions of anthropological archaeology). But this is most certainly a false dichotomy. Transhistorical regularities are only one way of framing an intellectual project that can articulate the ancient and the modern.

To resist the social evolutionary program need not entail consigning early complex polities to the dustbin of history or theory. The temporal distance that separates early complex polities from the modern can also be understood as providing a unique lens for viewing political life that lends our gaze a greater critical refinement through its profound historical depth and encompassing geographic breadth. Rather than compressing variation, investigations of early complex polities can revel in it, exposing the multiplicity of political strategies as well as antecedents of contemporary ambitions. This is not because the modern and the ancient are tied into a regular developmental chain but because the broad set of parameters that define current political life arose, were modified, set aside, and reestablished over a much longer time than can be encapsulated in the historically shallow term "the modern." Thus the articulation of modern and ancient can be framed around a historical anthropology of the political centered on a critical sociology of relations of civil authority. This frames the relationship as one not between theory givers (anthropologists) and data givers (archaeologists and epigraphers) but rather between complementary accounts of the complexities of political life engaged simultaneously with both theory and uniquely constituted records of authority (see Humphreys 1978: 22).

The political devices of the modern world only seem particularly clever and impenetrable when removed from history. For example, compare the following declarations:

> The earth was wilderness; nothing was built there; out of the river I built four canals, vineyards, and I planted the orchards, I accomplished many heroic deeds there. (Melikishvili 1960: #137)

> [T]he object lesson [of expansion is] that peace must be brought about in the world's waste spaces. (Beale 1956: 32)

The latter comes from no less an authority on imperialism than Theodore Roosevelt, whereas the former was inscribed by the Urartian king Argishti I who ruled an expanding empire in southwest Asia from approximately 785 to 756 B.C. (see chapter 4). Both rulers were speaking of "wildernesses" that had been occupied by other peoples for centuries; by reclassifying them as "waste spaces," expansion was not only conscionable, it was mandated. Roosevelt's strategy of redescribing imperialism as a triumph of order over the wild loses much of its grandeur, not to mention the force of originality, once exposed as a new gloss on a very old practice.[13]

Studies of early complex polities can thus inform investigations of the modern by assembling dialectical images—representations that juxtapose modern theoretical problems with ancient practices to illuminate novel approaches to repeated tropes of political representation. (On the dialectical image, see Buck-Morss 1989: esp. 67.) By casting representations of early complex polities as dialectical images, we allow the historical social sciences (to twist a phrase loved by journalists) to speak the past to power.[14] If histories of rupture allow the past, no matter how remote, to fade from view, consigned to a wholly incommensurable era, we lose the moral sway that comes from a broadened imagination of political possibilities. It is only through such an understanding of human politics beyond the modern that the imagination of alternative civil formations—what Prasenjit Duara (1996: 151–52) refers to as "other visions of political community"—can truly flourish.

A final motive for examining politics through the lens of early complex polities arises from the sense of recognition that their political histories evoke in modern observers. Unlike studies of early hunter-gatherer or small-scale agricultural communities that have traditionally framed their examinations as explorations of "the Other," investigations of early complex polities have been driven since their inception by a sense of recognition on the part of we moderns. In its crudest form, this recognition can take the form of classical elitism, as reflected in a remark Anthony Pow-

13. Pierre Bourdieu (1999: 57) made a similar claim, arguing that reflection upon the moments of initial formation (or "genesis") provide a powerful foundation for critique: "[B]y bringing back into view the conflicts and confrontations of the early beginnings and therefore all the discarded possibilities, [historical reconstruction] retrieves the possibilities that could have been (and still could be) otherwise."

14. Henry Glassie (1977: 29) has noted the utility of material culture in giving a voice to "the endless silent majority who did not leave us written projections of their minds," allowing archaeology to counter the "tales of viciousness" that provide the "myth[s] for the contemporary power structure." (See also Wylie 1999.)

ell attributes to Harold Macmillan, that "one would have no difficulty talking to Cicero 'if he came into Pratts' [an exclusive London gentlemen's club]" (Powell 1995: 37–38). However, in a more subtle form this recognition of the modern in the ancient can provide a sublime moral gravity to political reflection, as in Shelley's "Ozymandias." In that now-ruined king's command to "look on my works, ye Mighty, and despair!" we recognize in his pretensions the ultimate transience and incompleteness of our own civil orders. Such a humane motivation to look for the pale reflection of our own politics in that of ancient (yet not so far removed) social worlds can provoke a range of responses from optimistic assessments of the righteousness of current forms of civil order to pessimistic conclusions that the brutality of oppression has barely changed over 5,000 years. Both of these responses to early complex polities are productive in their own way insofar as they stimulate reflection on the nature of political authority in our lives. And here lies the illumination that archaeological studies can provide: by describing how dimensions of social life produced and reproduced political orders, the intersecting spatiality and historicity of our own world may become slightly less transparent and the State slightly less dim in our minds.

A Map of the Present Work

The Political Landscape is informally divided into two parts. The first (chapters 1 and 2) focuses on elaborating the central theoretical problems that arise from an effort to understand the constitution of authority through landscapes; the second part (chapters 3–6) examines the role of landscapes in four sets of political relationships describable in reference to the generic terms of geopolitics, polities, regimes, and institutions. These relationships are explored within three primary archaeological cases: the Classic period Maya; the early first millennium B.C. kingdom of Urartu, and southern Mesopotamia during the late third and early second millennia B.C.

Chapters 1 and 2 argue for two primary intellectual transformations in the contemporary theorization of political life. The first, taken up in chapter 1, centers on revising the conceptualization of space in the study of early complex polities. Although the dominant traditions in modern examinations of early complex polities define space as prior to the social world, such a philosophical stance relies on the highly problematic position of defending either an absolute or a subjective ontology that describes built environments as epiphenomena of historical process. Instead, I follow recent directions in geographic thought that describe space as emerging in *relations* between objects, an ontological revision that demands an account of landscapes as social artifacts that are produced and reproduced through varying dimensions of spatial practice.

Rather than considering spatial forms as dependent elements of temporal process (for example, the City as an accompaniment to the State), this notion of spatial practice highlights the ways in which spaces are created out of myriad social acts and actors ranging from the quotidian (such as a family building an addition to their home) to the extraordinary (such as the establishment of a polity's territorial boundaries through war or treaty). We can examine the production of landscapes along intertwined dimensions of spatial practice—in the bodily experience of spatial form, in the perceptual interaction of sense with place and aesthetic, and in the imagination of locale, world, and cosmos. No single dimension alone can give a sufficiently broad understanding of the spatiality of social life. Within a holistic vision of social space, landscapes can be understood as central elements in social production and reproduction. No longer conceivable as mere stages on which more deep rooted determinants unfold, the landscapes in which we live can be understood as instrumental in shaping the way we move through the routines and surprises of our daily lives, the affective responses engendered by places of particular meaning, and the ways in which we imagine our lives being reshaped. These are the ways we interact with landscapes and the sources of their political significance.

In chapter 2, I argue that, although the State has afforded us conceptual cover for typological debate and varied emplotments of metahistorical schema, it is a deeply flawed concept for an archaeology of politics. The suffocating focus on the evolution of the State has left the study of early complex polities without the theoretical apparatus for attending to the central problem of political analysis: what did early complex polities actually *do*? How did polities manufacture sovereignty? How did regimes secure power and legitimacy? How were subjects ordered? In developing models of political life so entirely consumed by links between time and types, we have been left with little idea as to how political practices produced and reproduced authority.

The absence of models for describing the operation of archaic States has led to a general diminishment of the import accorded politics vis-à-vis religious and economic arenas in archaeological reconstructions of social life (Conrad and Demarest 1984; Mann 1986; Van de Mieroop 1992; cf. Yoffee 1995) and to a reliance on the overly polarized concepts of coercion and consent to account for the solidarity of ancient political communities (Marcus 1993: 134; cf. Smith and Kohl 1994). As political scientists have long argued, the coalescence and administration of political communities cannot be adequately described in terms of simple pushes and pulls exerted by a governmental structure bent on subduing a recalcitrant population (Friedrich 1958; Gadamer 1975; Hobbes 1998; Oakeshott 1975). Rather, the central pillar of any political community is authority, the asymmetric, reciprocal public relationships where one actively practices a power to command that is confirmed by another as legitimate.

Chapters 3 through 6 examine the spatiality of political authority in early complex polities in reference to the four pivotal sets of relationships described above that in significant measure constitute civil life:

· the ties among polities that organize geopolitics;

· the links between subjects and regimes that create polities;

· the interaction of power elites and grassroots organizations that produce regimes; and

· the ties among institutions within a governing apparatus.

These four relationships are explored in reference to three archaeological landscapes:

· the Classic period lowland Maya (A.D. 250–900);

· the kingdom of Urartu (ca. 850–643 B.C.); and

· southern Mesopotamia from the Ur III period through the early Old Babylonian period (ca. 2125–1880 B.C.).

Geopolitics, the focus of chapter 3, refers to the formation of a political unit in space as coherent and distinct from neighboring polities. The central problem addressed in this chapter is the relationship among polities as they interact within a wider ecumene. Thus the pivotal concern of the chapter is to delineate spatial practices that shape the interpolity order.

Chapter 4 attends to the spatial dimensions of relationships that constitute polities, specifically the links between subjects and political authorities. The discussions in this chapter are concerned with the spatial production of internal coherence vital to the formation and routinization of authority. This includes not only the delineation of defined territories within which sovereignty is confined to the apparatus of a single authority, but also the creation of political identities—the association of identity with land through the demolition of prior commitments and the development of a constructed memory of landscape.

The concept of the regime, the focus of chapter 5, is used in this study to stand in for the host of implied structures and poorly articulated forms typically addressed under the rubric of urbanism. Comparative anthropology has shown that urbanism is not in itself a universal feature of complex polities; furthermore, there is such dramatic variation in city form within urbanized polities that it is truly impossible to speak of "*the* city" as a single historical space. By "regime" I mean the spaces defined by political and social elites with a direct interest in reproduction of structures of authority in concert with broader coalitions supporting authoritative rulers. Regime thus incorporates the spaces created both by the horizontal circuits of prestige, influence, and resources among elites and by the vertical ties (kin, ethnic, religious) that extend down to grassroots levels. As a number of recent studies have suggested, many of the places that we define as central to urban environments arise out of the practices of such regimes (see, for example, Elkin 1987; Stoker 1996; Stone 1989). Furthermore, it is in the context of regimes that we can explore the aestheticization of a political apparatus through the sensual dimensions of formal perception (such as the evocative potency of urban experience) and the imaginative dimensions of representational media (such as the attachment of values of civility and humanity to urban life).

Temple, palace, and market have long served as proxy terms for institutional complexes based on sacred, political, and economic power. However, the actual spaces have rarely been described as fundamental to institutional operation. Chapter 6 focuses on both the production of physical institutional spaces and tensions between rival imaginings of political legitimacy. The built spaces of political institutions have often (particularly in the Classical world) been hailed as triumphs of human architectural genius expressive of developing human creativity (see, for example, Lloyd 1980: 12–13). However, they were more profoundly key elements in the production of political authority, enabling regimes to regularize the demands placed on subjects. But despite the impression of coherence that regimes foster, institutions can also provide prominent sites of factional competition. Thus, this chapter also addresses the spatial dimensions of institutional fragmentation that can promote crisis, fragmentation, and collapse.

The final chapter provides a concluding formulation of the central themes of the book and contextualizes these discussions in reference to three primary issues: the role of a constellatory analysis in a comparative archaeological program, the relationship between ancient landscapes and contemporary politics, and the status of the early complex polity as an object of analysis. These discussions are intended to point future theory and research in potentially productive new directions.

Taken as a whole, *The Political Landscape* is an attempt to fill in a gap between contemporary theorizations of civil life and investigations of politics in the ancient world. I try to find a balance between attention to general theory, on the one hand, and a commitment to the details of archaeological and epigraphic explication, on the other. The archaeological perspective offered here is not assembled out of a regionally focused study of a single case or a traditional effort at cross-cultural comparison. Instead, these archaeological studies outline constellations of intersecting political practices. I adopted this constellatory approach to the material for two reasons. First, due to the exigencies of preservation and traditions of research, no one locale presents useful cogent evidence for all dimensions of the political landscape. Thus, practical considerations demanded that the discussions range beyond any single case study. Second, the book is intended to help resuscitate a genre of anthropological writing that explores material in a comparative spirit without yielding to the reductionist tendencies that tend to cripple many such works. Thus, it was critical that each case be allowed to develop in its own right without the compression that results from traditional comparison.

It is with some trepidation that I step away from the primary region of my own field research in the Caucasus. However, to refuse to move our theory-building beyond single emblematic regional cases threatens to rest claims to interpretive priority on the unsatisfactory grounds of experience rather than argumentation. Indeed, hesitancy to move beyond the single case threatens to balkanize research programs that, within an anthropological archaeology, should flow quite freely into one another. Thus the cases here are juxtaposed in order to tease out the central argument — that polities in early complex societies operated through landscapes — without suggesting that these practices have any logic beyond the historical constellations of power and legitimacy constituting authority. I have restricted the comparative horizon of the present study to early complex societies only in order to provide for a closely delineated frame of reference. I do not mean to restrict political landscapes to the narrow set of cases discussed here, just as I do not view the limited group of political relationships at issue in subsequent chapters to preclude others. These steps have been taken merely to focus an investigation that otherwise might have easily grown to several volumes.

In setting forth these caveats, I must also note that the importance of landscape to political analysis should be neither over-inflated nor under-stated. To attend to political landscapes is not to suggest that landscapes are exclusively or even primarily political or that politics is exclusively about landscapes. Neither exhausts the other. Yet clearly the relationship between the two constitutes a critical problem for the social sciences—for how can we begin to understand, for example, the emerging nations of the Commonwealth of Independent States if we fail to recognize the problems raised by the uneven distribution of military and industrial resources and the territoriality of Soviet republics? Alternatively, can there be an understanding of the conflicts between Israel and the Palestinian Authority in the absence of a theory of how political authority comes to be so tightly wed to specific parcels of land? Can we begin to approach an account of the French Revolution that fails to apprehend the Bastille and the barricades? The remainder of this book is an attempt to examine the links between space, time, and political authority in order to demonstrate that it is impossible to describe political authorities independent of the landscapes they created: the regions they united, the cities they built, the inspirations evoked in the monuments they raised, and the cartographic desires they inflamed.

CHAPTER I

Sublimated Spaces

Space is the mark of new history and the measure of work now afoot is the depth of the perception of space.

Charles Olson, "Notes for the
Proposition: Man Is Prospective"

Charles Olson, poet and precocious postmodernist, was wrong. Although his call in 1948 to spatialize our understanding of the human past is regularly trotted out as an intellectual precursor to late-twentieth-century trends in social and literary theory, it would be difficult to argue at present that space has indeed become central to historical reflection. This is particularly the case for investigations of early complex polities—ancient political formations in which authority was predicated on radical social inequality, legitimated in reference to enduring representations of order, and vested in robust institutions of centralized governance. Despite halting movements toward geographic critiques of modernity, the vast horizon of human experience beyond the reach of the modern remains without a clearly theorized sense of spatiality. Just one year after Olson issued his exhortation, the publication of Leslie White's *The Science of Culture* signaled a profound intellectual move within anthropology toward neo-evolutionary accounts of human history that hinged on sublimating spatial difference. The goal of this enterprise was to establish the foundational temporal currents within human cultural development (what Robert Wright [2000] has termed "the logic of human destiny"). Only in the past decade has this sublimated sense of space encountered philosophical resistance in the form of a neo-historicist renaissance built on primarily phenomenological intellectual foundations. A steadily increasing number of archaeological studies have called for a more active understanding of space, one centered on the symbolic content of places and the meaningfulness of landscapes. However, the pronounced tendency of neo-historicists to theorize places as socially active yet analyze them as pale reflections of culture, identity, or a universal humanity suggests that exactly how this relationship between place, time, and social life is to be understood remains an incomplete project. Although Olson's assessment today still seems optimistic, a handful of critical works from the past 25 years of archaeological research have led the way toward an understanding of space situated within a historical account of social life.

In this chapter, I sketch the outlines of a spatial approach to early complex polities, advancing a critique of evolutionary and historicist philosophies of space and arguing for a relational ontology of human landscapes. By foregrounding the production of landscapes within a broad set of spatial practices that include the experience, perception, and imagination of space, I allow politics to emerge from the long shadow cast by formal typologies and historical schema to set the analytical agenda squarely on the constitution of authority. Although social evolutionary thought has recently come under sustained criticism throughout the social sciences, it remains the dominant historiographic model for representing the formation of early complex societies. Yet the critical dissent that has effectively marginalized social evolutionary theory across much of the social sciences has yet to produce a single theoretically encompassing archaeological treatise. The same might also be said for the neo-historicists who originally rose to prominence under the banner of the "post-processual" movement in archaeology during the 1980s and 1990s. While there has been extensive critical engagement with post-processualism as an intellectual project, scrutiny has not extended to the phenomenological turn in studies of ancient landscapes. It is therefore important that we undertake an examination of the status of space within contemporary theorizations of early complex societies as a point of entry for an account of political landscapes. I hope that the reader will forgive the necessarily didactic portions of the following discussion. These are brought forward here in order to create the intellectual space within which the more theoretically prognostic aspects of the current project can develop.

It is worth reiterating at the outset the denotative distinction that I draw between landscape, space, and place (outlined in the introduction) because these terms have provoked repeated attempts at close definition over the past two decades. However, rather than wade into these disputes, I want to provide compact definitional statements for each term that will serve to organize the following discussions. Space has become a focus for the ire of some phenomenologists, such as Edward Casey (1997: 333–35), who regard it as a hostile, rationalized, infinite container that has come to encroach on traditional, meaningful, and highly localized senses of place. However, such accounts leave us with both a highly unproductive understanding of space and an overly sentimental account of place, neither of which provide particularly robust conceptual platforms for analytical work. Thus I prefer to describe space as that philosophical rubric under which all problems related to extension and the parameters of synchronic relation may be discussed. Place, a far more tailored concept and thus of considerably greater utility within social science research, refers to specific locations invested with meanings that arise from a diachronic sense of their sociocultural instantiation. That is, places emerge within specific histories. But, whereas places tend to be rather restricted in extent, landscapes are far more embracing, both spatially and temporally, encompassing not only specific places and moments but also the stretches between them: physical, aesthetic, and representational. In other words, landscapes assemble places to present more broadly coherent visions of the world. As visions, however, they are ultimately rooted in specific perspectives that advance particular ways of seeing, of living, and of understanding.

Charles Olson provides an appropriate bellwether for this study because he was not only attuned to the spatial dimensions of modern life but also captivated by investigations of the ancient world.[1] Although his perspectives on the past were plagued by a ponderous mysticism, Olson recognized in archaeology a unique capacity to deepen *and* broaden human knowledge. To their detriment, he argued, modernist examinations of the ancient world had reduced early complex polities to overly generalized forms devoid of content—bland sociological types ripped from the spaces that make these bygone ways of life uniquely compelling. The promise of archaeology lay, he argued, in its unique ability to create a new history situated in and developed out of space rather than removed from

---

1. Clark 1991: 197. Olson organized a series of lectures at Black Mountain College in 1953 that attempted to bring together archaeology and mythology toward what he envisioned as an exploration of the archaic roots for a new "mythological present." Representing archaeology on the panel of lecturers was Robert Braidwood of the University of Chicago Department of Anthropology (ibid., 233).

it. Such an intellectual course, as we shall see, cuts against the grain of the traditions that have dominated the study of early complex polities since the late nineteenth century.

Social Evolutionism and Absolute Space

Social evolutionism describes a prospect on human history that visualizes an overall shape to human social development, a progress toward increasing complexity that can be explained in reference to a set of rational determinants. Although sharp disagreements as to the exact contours of evolutionary pathways (unilinear, multilinear, parallel, convergent, divergent) and the mechanism(s) that drives transformation (from ecological to cultural determinists) separate social evolutionists into various schools of thought, several key points provide the common foundation to the social evolutionary view. First, there is a necessity to social evolution that propels us from simple forms of association to more complex societies that are larger in scale and more differentiated in internal structure. Although generally stripped of the evaluative sense of the term "progress," social evolution remains essentially teleological in its emplotment of history. Second, social evolution may move at a different rate in different parts of the world, but the shape and mechanism are universal. That is, social evolution is a global process rooted in the fundamental nature of human society and so is ultimately independent of both spatial variation and human action.[2] Third, priority in the determinants of social transformation is ceded to material dimensions of life—adaptation, relations of production, demography—that then shape the "non-recurrent" particulars of belief, thought, and performance (Steward 1972: 209). This allows social evolutionary analysis to focus on the rise and fall of a handful of societal types, conjoined through their essential determination in the material conditions of existence, despite the wide variability in cultural expressions.

With its intellectual roots as firmly planted in the brutish conservatism of Herbert Spencer as in the transcendent radicalism of Karl Marx, the

---

2. Only by denying, or sharply circumscribing, the ability of human agents to actively shape the course of history—to redirect or forestall its ultimate logic—can social evolutionism aspire to be a liberative theory, as described in Trigger 1998. Thus, paradoxically, the hope for human freedom (within an evolutionist framework) rests on enslavement to an overriding metahistory.

argument over social evolution cannot be conducted through a political litmus test, as a number of writers have suggested. To position social evolutionism as the sole beacon of hope for a better world against the regressive relativism of postmodernism, as Bruce Trigger (1998: xi–xii) has suggested, is to forget evolutionism's complicity in building the global apparatus of industrial exploitation and to suggest that all humanity's ambitions yearn for the single historical trajectory defined by Western social thought. Such an argument unwisely transforms social evolution from a theory of history built on contestable facts to a religion of history based on incontrovertible faith—hence the stridency and emphasis on orthodoxy among many believers. But to dismiss social evolution based on its deplorable role in providing a racist foundation for colonialism—as Anthony Giddens (1984: 236–38) seems to do—is to forget its deployment, generally in Marxist form, in a host of twentieth-century revolutions that overturned colonial relationships. Social evolutionism has considerable blood on its hands, yet that of the oppressor mixes with that of the oppressed. Thus, the critique of social evolution must be leveled on philosophical, rather than political, grounds. Numerous critics have probed the weaknesses in social evolutionism's historical view (for example, Diamond 1974; Rowlands 1989; Shennan 1993; Yoffee 1993), but my discussion here is concerned solely with drawing out its implicit spatiotemporal ontology and forwarding a critique of its deployment in the study of early complex polities.

Although social evolutionism is, above all, a theory of time, of the shape, pace, and direction of history, its foundational conceit—that world history may be understood under the rubric of a unified law of social change—is predicated on the reduction of space to a social constant. That is, variations in space must be insulated from affecting social transformations so that explanatory power may be vested exclusively in the temporal axis. If space were to hold the potential to shape the course of future transformations, then temporal variation would be difficult to define in universal terms. Space, within an evolutionary approach to social life, must be described as an absolute.

To illustrate the dependence of social evolutionary thought on an absolute ontology of space,[3] we might consider the rise of complex polities in fourth millennium B.C. southern Mesopotamia. This profound historical transformation was marked by broad shifts in the spatial organi-

---

3. In using the term "ontology" I am referring to the description (generally implicit) of the nature of things that underlies a given theory or system of thought.

zation of social life, including altered settlement patterns, the advent of urbanism, the dawn of new monumental and vernacular architectural forms, and increasingly politicized landscape aesthetics. But with space held constant as an explanatory variable, the cities of southern Mesopotamia can only be understood as passive expressions of evolutionary process; they cannot be ceded any role in shaping the course or content of that process. It is only as a result of rendering space epiphenomenal to deeper temporal process that the appearance of urban settlements and monumental architecture in the Indus valley in the third millennium B.C., in northern China during the second millennium B.C., and in lowland Mesoamerica by the end of the first millennium B.C. can all be articulated with the transformations in fourth millennium B.C. southern Mesopotamia into a single history describing the evolution of the State. Where time is flux and causation, space is absolute and inert.

The philosophical premises of this absolute ontology of space were most famously outlined by Isaac Newton: "Absolute space, in its own nature, without relation to anything external, remains similar and immovable. Absolute space is the sensorium of God" (quoted in Garber 1995: 302). By declaring space to be prior to experience, Newton transformed it from a variable dimension of existence (as in the Aristotelian tradition) into an object in itself. According to Newton, even if the universe were devoid of matter, space, as a three-dimensional structure, would still exist (Sklar 1974: 161). It is this account of space that underlies William Blake's famous sketch of the "Ancient of Days" where the deity, assuming the role of master cartographer, reaches down from heaven to take the measure of the universe. Newtonian space is a unique sort of object, one that is unchanging over time and empirically incomprehensible; space can only be inferred from observable phenomena. Matter exists within this container we call space in the sense that objects coincide with a set of preexisting, permanent points (Sklar 1974: 162). Space is thus independent of its occupants and, in its fundamental nature, unaffected by them.

Although the absolutist ontology is quite clear that changes in relationships between objects hold no implications for the shape of space itself, what is unclear in Newton's account is the degree to which space can influence the objects that inhabit it. This uncertainty defines an important theoretical split in the absolutist position as it has been operationalized within social evolutionary approaches to early complex polities. We can outline two primary variations on spatial absolutism within the social evolutionist tradition. The first, developed out of a seventeenth-century mechanical metaphor, describes the social world as a machine that must be understood in geometric terms.[4] This mechanical absolutist position holds that space has no effect on historical process and, as a consequence, spatial variables are largely irrelevant to historical explanation. The goal of spatial analysis, within this mechanical tradition, is to shed light on the fundamental geometry that structures the world. By this account, the general process of social evolution plays itself out in an entirely undifferentiated space. The second absolutist position arises from a metaphor employed by eighteenth- and nineteenth-century natural scientists such as Georges-Louis Buffon, Jean-Baptiste Lamarck, and Charles Darwin who described the world as an organism.[5] Detailing the workings of this organism involved explication of nature as a set of functionally interrelated parts. The role of analyses of space within an organic absolutist ontology is to define determinative processes organizing the spatial relationships between components. Foremost among these organic determinative processes in the twentieth century has been adaptation.

MECHANICAL ABSOLUTISM

The roots of a mechanical account of absolute space appear in the cartographic traditions of Western Europe by the eighteenth century, rising to prominence in the geographic writings of Carl Ritter and Alexander von Humboldt. Humboldt's ambition to "discern physical phenomena in their widest mutual connection, and to comprehend Nature as a whole animated and moved by inward forces" (1847: xviii)[6] blended the emerging geographic positivism with the lingering romanticism of a divine Nature. Ritter similarly sought to describe an essential unity behind geographic variation, but a synthesis of regional geography with a cosmic sense of purpose rested more squarely in his writings on a historical teleology. Geography, in Ritter's view, entailed an exploration not just of the shape of the world today but also of the "unseen hand" that charted its historical course (Ritter 1874: xiv–xvi; see also Cannon 1978: 105; Peet 1998: 11).

The foundations of social evolutionist accounts of early complex poli-

---

4. Foremost in constructing this metaphor were Galileo, Descartes, and Newton himself (see Glacken 1967: 391).

5. The roots of this metaphor can be found in Classical sources (Glacken 1967: 134; Kahn 1960: 220–30).

6. I refer here to the more abstract, scientific threads that run through Humboldt's work rather than to the more humanistic elements that, as David Harvey (2000: 547–49) points out, owed a profound debt to Kant's *Geography*.

ties, as elaborated at the close of the nineteenth century by Lewis Henry Morgan, rest squarely on Ritter's combination of spatial positivism and historical teleology. Morgan employed comparative ethnography to define three basic social forms—savagery, barbarism, and civilization (each with numerous subphases)—distinguished from each other by technological criteria (see Morgan 1985). These forms of contemporary societies were in fact, Morgan argued, sequential stages of human development that all human societies have progressed through to relative degrees: "Like the successive geological formations, the tribes of mankind may be arranged, according to their relative conditions, into successive strata. When thus arranged they reveal with some degree of certainty the entire range of human progress from savagery to civilization" (quoted in Strong 1953: 389). The primary geographic implication of Morgan's evolutionism was that the shape of human history could henceforth be considered independent of place. No matter where a society was located or what its configuration, its history could be fit within a universal schema. Freed from the epistemological constraints imposed by geographic difference, Morgan was thus able to study contemporary peoples who were related to him spatially—contemporaries tied to the expanding European world through a colonial geography—as "survivals" from the past more profoundly related to him temporally. Thus the Iroquois, to use one of Morgan's case studies, are not understood as occupying a unique place (and thus possessing a unique history) but rather as holdouts from a universal developmental stage now long past for Euro-American societies.

Although space was inconsequential to explanation, it was useful as an expressive feature of evolutionary logic; dimensions of spatial form, such as architecture, held analytic significance as they marked each of the stages: "House architecture . . . affords a tolerably complete illustration of progress from savagery to civilization. Its growth can be traced from the hut of the savage, through the communal houses of the barbarians, to the house of the single family of civilized nations" (Morgan 1985: 6). Thus, built form is highly expressive of evolutionary process, of both the shape and teleology of history. But the form of, for example, the "communal house of the barbarians" can in no way affect the ultimate realization of the historical drama in the rise of the single family house or, to update Morgan, of the split-level suburban ranch-style home so emblematic of postwar America.[7]

---

7. Here we might note the very different colonial discourses on the power of redesigned domestic architecture to transform local societies (Comaroff and Comaroff 1997: 274–322).

Morgan's representation of simultaneity as sequence, though strongly criticized during the early twentieth century, experienced a revival in the 1960s led by anthropologists such as Marshall Sahlins, Elman Service, and Morton Fried who explored the implications of the social evolutionary revival sparked by Leslie White and Julian Steward. This second generation of neo-evolutionists shared with Morgan an emphasis on the creation of typologies of social forms based on living groups and their projection onto a historical sequence as homologues of prehistoric and early historic societies visible in the archaeological record. Sahlins and Service (1960: 37) developed the most widely cited evolutionary typology: a four-staged sequence that began with bands of hunter-gatherers and progressed through tribes and big-men societies (which they termed chiefdoms) to culminate in archaic civilizations and industrialized nation-states. Following Morgan, they read differences in cultural practices and social institutions across space—contemporary societies distributed around the globe—as reflections of differences in historical development along a generalizable temporal scale. Thus our neighbors, linked to us across space, were reconfigured as, in Service's oxymoronic phrase, "our contemporary ancestors" (1975: 18).

Although mechanical absolutism provided neo-evolutionism with its foundational teleology and faith in universal laws, it was the quantitative turn in geography during the mid-twentieth century that provided a set of highly developed analytical techniques for articulating dimensions of form with the spatial logic of history. The strand of spatial positivism that was to exert the most profound effect on the archaeology of early complex polities was the set of procedures that clustered around the study of settlement location. Reacting to Richard Hartshorne's interest in geography as regional history, Fred Schaeffer issued a call in 1953 for a more rigorous methodology that might assist in formulating general laws of spatial relations (Schaeffer 1953: 226; see also Hartshorne 1939). Such methods for distilling the complexity of spatial relations into simplified representations had been developing both within and outside geography decades before Schaeffer's exhortation. In the late nineteenth century, J. H. von Thünen used simple "isotropic plains" to describe the distribution of agricultural land around cities in purely geometric terms; Alfred Weber employed a similar technique to describe the distribution of industrial features; Felix Auerbach described the "rank-size rule," which held that in industrialized nations the largest city hosts twice the population of the second ranked city, three times that of the third ranked, and so forth; and Walter Christaller used idealized hexagons to describe hierarchies in

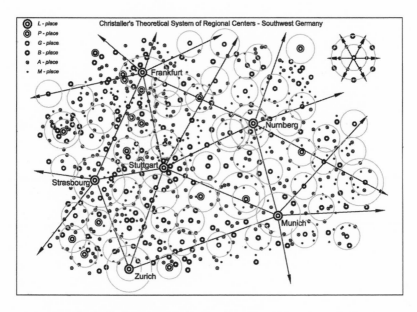

FIGURE 4. Christaller's geometric lattice of settlement location: idealized locational networks in southwestern Germany. (Redrawn from Christaller and Baskin 1966.)

the distribution of cities in southern Germany (Auerbach 1913; Christaller and Baskin 1966; Thünen 1966; Weber 1957).

Spatial regularity, Christaller famously argued, arises from universal physical principles. Describing these principles became a priority for quantitative geography in the 1950s and 1960s: "The crystallization of mass about a nucleus is part of the elementary order of things. Centralistic principles are similarly basic to human community life. In this sense the town is the center of a regional community and the mediator of that community's commerce; it functions then as the central place of the community" (Berry and Pred 1965: 15). Thus, Christaller was able to extrapolate regularities in the distribution of similarly sized cities in southwestern Germany into an idealized geometry of settlement distribution (fig. 4). The central place theorists, the first wave of locational geography, forged a grand synthesis of geographic positivism and cartographic empiricism to express notions of spatial regularity; however, the determinants of "centralistic principles" in human settlement were highly undertheorized. In Christaller's account, the interactions of place and market seem so peculiar to pre-war southwestern Germany that the potential for generalization is by no means clear. By marrying central place patterns to neo-classical economics, a second wave of locational geographers, led most notably by August Lösch (1954) and Walter Isard (1956), generalized models of spatial regularity into a geometry of human behavior predicated on assumptions of market rationality and efficiency.

Location theory provided not only the tools for detecting and explaining patterns that underlay extant spatial systems but also the proper rules for regional planning that would bring settlement systems into accord with what John Q. Stewart termed the mathematical rules for human behavior that constituted a new "social physics" (1947: 179). In the attachment of idealized models of spatial geometry to a theory of economic (market) determinants, the mechanical ontology of space reached its logical culmination, establishing spatial distributions as reflective of a universal, unchanging set of rules. This was a vision of Newton's "sensorium of God" well suited to the mid-twentieth-century expansion of American-style capitalism.

Four decades after Christaller's original work (and just as location theory came under heavy assault within geography; see, for example, Harvey 1973; Lowenthal 1961; Massey 1973), archaeologists turned to location theory in order to frame neo-evolutionary studies of early complex polities. What location theory offered was a set of methods that might be operationalized archaeologically to describe the appearance of social evolutionary stages.[8] In the hands of archaeologists, the categories of social forms defined by neo-evolutionists quickly ceased to be regarded as ideal types. Rather, they came to be (for many researchers) more real than the actual societies that produced the remains discovered through archaeological survey or excavation. As a result, the theoretical stakes of investigation no longer centered on close reconstruction of past ways of life but rather on articulating social groups with social evolutionary categories, compressing variation in order to fit the restricted set of types.[9] Great the-

8. This was particularly a problem for the burgeoning field of archaeological survey, which was based on the examination of surface materials across a broad region (in contrast to the traditional method of excavation, which examines sequential subsurface levels of occupation in a single locale).

9. Despite protests from a number of social evolutionists (Sanderson 1995; Trigger 1998), this critique leveled by Giddens (1984: 236–43) remains a valid point. The very poetics of social evolutionary writing (e.g., Johnson and Earle 1987) belie the centrality of variation compression to social evolutionary thought in that a set of case studies (both ethnographic and archaeological) are described in order to illustrate a categorical totality, such as the Simple Chiefdom or the Archaic State.

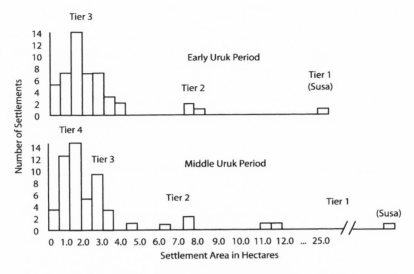

FIGURE 5. A site size histogram for two eras of settlement on the Susiana plain of southwestern Iran. (Redrawn from Johnson 1973.)

oretical consequences rested on the development of analytical tools for categorizing archaeologically known societies according to social evolutionary categories. Yet how was a universal history to be reconciled with the spatial particularity of the archaeological record?

By importing the basic features of locational approaches, Gregory Johnson (1973) proposed an ingenious solution to this problem that he described in reference to his field research on the emergence of the State during the fourth millennium B.C. in the Susiana plain of southwestern Iran. Johnson suggested that variation in site surface area within a region corresponds directly to hierarchical levels of decision making that organize local exchange within polities. The payoff of Johnson's approach was that, by measuring the areal extent of sites, plotting them in simple histograms, and looking for breaks in scalar distribution, archaeologists had a spatial proxy measure for stages of social evolution (fig. 5). Band-level societies, with only one level of decision making, had only one size of settlement. Tribal societies had a few larger sites that served as special meeting centers. Chiefdoms had three hierarchical levels of settlement sizes, and the State, as the most complex and highly integrated social form, had four or more hierarchical levels of site sizes. Thus, Johnson could track in the Susiana plain settlement patterns the transformation of a two-level (tribal) settlement hierarchy (small centers and villages) during the early fourth millennium B.C. (terminal Susa A period), into a three-tiered hierarchy in the succeeding early Uruk period (ca. 3750–3500 B.C.), and finally into a four-tiered (State) hierarchy by the middle Uruk period (ca. 3500–3200 B.C.). Regularities in spatial patterning were intelligible as expressions of the underlying mechanics of social evolution.

Locational approaches to the evolution of complex polities have proliferated since the 1970s to include rank-size approaches, applications of Thiessen polygonal lattices, and central place modeling. Studies utilizing a locational approach have appeared in archaeological contexts around the globe, including the Classic period Maya, the Aztec Empire, Iron Age Europe, and Late Imperial China (see, for example, Ball and Taschek 1991; Falconer and Savage 1995; Hammond 1972; Inomata and Aoyama 1996; Lamberg-Karlovsky 1989; Skinner 1977). What unites these varied approaches is a shared mechanical spatial ontology, a commitment to explaining regularities and variation in spatial patterns in terms of a universal geometry of settlement determined, in the last instance, by the logic of social evolutionary process. This evolutionary process holds no import for the fundamental nature of the spatial logic described by location theory. By articulating a universal historical process of social evolution with a universal spatial mechanics through meditating assumptions regarding the economic determinants of behavior, space is given independence from the effects of the social world (behavioral motivations are universal and insulated from sociopolitical variation) and concomitantly precluded from exerting any influence on future directions of social change.

Although mechanical absolutism has traditionally focused on analyses of regional spatial patterning, Kent Flannery (1998) has extended this perspective to incorporate comparative analyses of public buildings and mortuary architecture. Locating complexity in both social stratification and a specialized priesthood, Flannery suggests that monumental palaces, special residential quarters for priests, and profound differentiation in the quality and extent of tombs can serve as markers, or "rules of thumb," of archaic States. Although archaeologists have long used similar criteria for describing complexity (for example, Childe 1936), what is notable about Flannery's contribution is his formalization of these built features into indexes of the transition between two social evolutionary forms, the Chiefdom and the State. Flannery's analysis rests securely within the mechanical absolutist tradition precisely because the built features he identifies as indicative of formal social types are described as arising (albeit with some room for variability) as a result of the inherent dimensions of the evolutionary transition between State and Chiefdom. In other words, neither palace nor royal tomb nor priestly residence plays an active role in forwarding social transformations. They are, instead, geometrical forms that accompany movement through social evolutionary stages, superficial proxies reflective of social transformations but insulated from them by the determining temporality of evolutionary history.

Mechanical absolutist accounts of early complex polities articulate closely with what Service (1978: 27) refers to as "integrative theories" of political formation. By emphasizing the systematicity of regional settlement distributions—the spatial logic of distributions, architectural forms, and so forth—mechanical absolutism presumes a centrifugal account of political systems where well-ordered polities expressed in well-ordered landscapes achieve a level of social consent that results in stability and coherence. In this sense, the emergence of the State is conceived of as a realization of a logic of both space and time. This description of early complex polities may be objected to on a number of grounds, from the evidentiary to the theoretical. Perhaps the most powerful of these critiques suggests that, in their tendency to find regularity in settlement distribution, locational approaches have the effect of an ex post facto legitimation of political authority, dismissing the vagaries of power, domination, and hegemony under the banner of a naïve contractarianism. Though a rather cynical criticism, rooted in the fact that few locational analyses arrive at the conclusion that a given case did not operate within the parameters given by neo-classical economics, it is nonetheless the case that, by predicating political consolidation on a spatialized correlate to consent, the actual means by which consent might be manufactured are obliterated by an overdetermined formalism.

A corollary criticism of the mechanical absolutist ontology centers on its hard-wired materialism. As a case in point, a recent application of central place models to Classic Maya polities in Honduras concludes that "the Classic Maya do not appear to have developed a strong state ideology or effective administrative systems that would have overcome technological and economic factors" (Inomata and Aoyama 1996: 306). But the discovery of ultimate causation in tendencies to cost minimization is profoundly tautological because these assumptions lie at the heart of the locational model. That is, location theory assumes the primacy of economic factors in decision making. To suggest that analysis can then prove the primacy of economic factors in locational choices is merely to restate the preliminary assumptions of the model and says nothing about the political milieu within which cost minimization or other economic bases for decision making might have been forwarded as significant priorities. It is

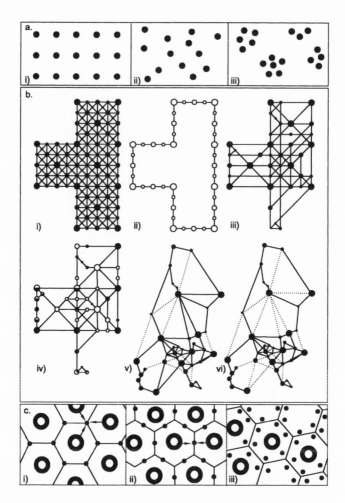

FIGURE 6. The representational aesthetics of central place
theory. **a.** Patterns of points: (i) regularly spaced; (ii) random
scatter; (iii) cluster. (Redrawn from Hodder and Orton 1976:
fig. 3.1.) **b.** Johnson's proposed settlement size and location
lattice showing: (i) the model lattice; (ii) sites that might be
smaller than predicted due to truncated complementary region;
(iii) observed site locations relative to model; (iv) deviation of
observed site sizes from predicted; (v) deviations in relative
site to site distances; (vi) derived proposed settlement lattice.
(Redrawn from Johnson 1972.) **c.** Idealized distribution of
centers according to Christaller's (i) market, (ii) transport,
and (iii) administrative principles. (Redrawn from Hodder
and Orton 1976: fig. 4.5.)

thus unclear what work such analyses can do in elucidating how early complex polities operated.

On a more ontological level, the mechanical absolutist perspective may be faulted for insulating space from the social. Space is instead merely descriptive of more deeply rooted imperatives, be they temporal (as demanded by evolution) or behavioral (as demanded by neo-classical economics). The practical result of the mechanical ontology is the displacement of spatial analysis from real places onto an idealized abstracted geometric plane. In the case of central place theory, only two spatial dimensions are considered relevant: distance and size. This is made quite clear in the spatial diagrams that accompany most locational analyses (such as those in fig. 6) where the particularities of geography and environment are erased, leaving only differentially sized dots on an undifferentiated background. The argument of mechanical absolutism is that this organization of dots represents no place but every space. Local geography is irrelevant to an explication of the fundamental workings of space except as it accounts for (minor) deviations from the fundamental rules. The result, of course, is that spatial analyses derived from a mechanical absolutist position tell us very little about any place even as they aspire to describe everywhere.

A final objection to mechanical absolutism, as it conjoins with social evolution, is the basis for its singular privileging of the temporal in the explanation of social phenomena. There is no clear justification for redescribing what are manifestly spatial relationships among contemporary societies as temporal links. Eric Wolf has made a compelling argument against the compression of space into genealogy because it unjustifiably depicts social groups as "precursor[s] of the final apotheosis and not a manifold of social and cultural processes at work in their own time and place" (1982: 5). The subservience of space to time tends to reduce spatial analyses to purely static frameworks dependent on one-to-one correlations between stages and forms.

ORGANIC ABSOLUTISM

The second form of spatial absolutism current in investigations of early complex polities, a position predicated on an organic ontology, developed in response to the marginalization of local environments within the mechanical geometries of classic social evolutionary theory. Organic absolutism replaces the abstract geometry of spatial laws that marks the mechanical ontology with the concrete biological demands of the natural world, emphasizing the determinative capacities of the environment on social evolutionary process. What unites the organic and the mechanical traditions, as expressed in evolutionary examinations of early complex polities, is a commitment to insulating objective space (as geometric or natural law) from the effects of the social; what divides the two positions is that whereas mechanical laws, such as those described by location theory, have no effect on universal historical process, the laws of organic space, mediated through regional and local environments, do affect the course, pace, and process of social evolutionary change.

We can see the organic position deployed in two primary modes. The first derives from applying ontogenic descriptions to social history. In *Decline of the West,* Oswald Spengler (1932) described civilizations as organisms that followed the inevitable life cycle of birth, maturity, and decay. In Spengler's sense, social phylogeny recapitulated ontogeny. David Clarke (1968) elaborated this framework into a "culture system ontogeny" where evolution moved through stages from birth to death—stages he termed formative (florescent), coherent (classic climax), and postcoherent (postclassic). Although of rather restricted theoretical import, this developmental sense of organic social evolution has proven to be enduring not only in the discipline of history, where it reappeared in Paul Kennedy's (1987) ontogenic account of the rise and fall of the "Great Powers" since A.D. 1600 (see chapter 3), but also in archaeology where ontogenic terminology is often used to describe regional cultural sequences (such as the formative [pre-Classic], Classic, and post-Classic Maya). However, the ontogenic form of organicism has had only a minor impact on the analysis of space, place, and landscape.[10]

The second, or environmentalist, mode of organic absolutism has had a more pervasive influence on archaeology's theorization of space and time. Rooted in cultural ecology, this perspective emphasizes the capacity of the natural world to shape and constrain social evolution. The environmentalist position traditionally traces its descent to the naturalists Lamarck, Darwin, and Charles Lyell. Leon Batista Alberti (1988) provided a metaphorical foundation for the organic position, employing a corporeal simile in his fifteenth-century monograph *De Re Aedificatoria* that

---

10. A recent examination of the growth of Grasshopper Pueblo in the American Southwest draws an explicit connection between the life cycles of organisms and settlement architecture, ostensibly to provide a foundation for modeling the effects of ethnic migration and community growth (Riggs 1999). However, the analogy does not in this context do any real analytical work, serving only as a metaphor for the founding, growth, and abandonment of the settlement.

described parts of buildings corresponding to each other in the manner of parts of the body. And organicism's deep-rooted teleology was explicitly formulated by the early nineteenth century when Aloys Hirt wrote: "One may consider every work of architecture as an organic whole, consisting of primary, secondary and contingent parts, which stand in a definite volumetric relationship to each other. In the case of organic bodies nature herself has determined the relations of the parts to each other in accordance with individual ends" (1801: 13, translation from Eck 1994). Space is thus determined not by mechanical geometric laws but by organic laws of nature.

The roots of modern organic absolutism lie in the application of nineteenth-century naturalist thought to explanations of human history. In an essay written in 1904, Halford Mackinder set forth a highly determinist account of the impact of global geography on world history: "[I]n the present decade we are for the first time in a position to attempt, with some degree of completeness, a correlation between the larger geographical and historical generalizations. For the first time we can perceive something of the real proportion of features and events on the stage of the whole world, and may seek a formula which shall express certain aspects, at any rate, of geographical causation in universal history" (1904: 422).

According to Mackinder, the geographical "pivot" of (Old) world history is centered on the Eurasian steppe, a vast highway at the heart of the continent linking east Asia, Europe, and South Asia. The importance of this region lies in the ability of the resident group to open and close access to disparate areas of the continent. Mackinder shared the absolutist understanding that changes in the social realm driven by "universal history" have no impact on the fundamental nature of space. But Mackinder departed from the mechanical absolutists in suggesting that elements of natural geography—topography, climate, hydrology, and other environmental variables—play a causal, even determinative, role in universal history.[11] Although the geopolitical conclusions derived from the analyses of Mackinder and other geographic determinists (such as Friedrich Ratzel or Ellen Semple) are of little consequence today, their legacy lies in what David Livingstone has referred to as their "geographical experiment . . . a manoeuvre designed to hold together the natural and social worlds under one explanatory umbrella . . . [using the vocabulary] provided by the Neo-Lamarckian construal of evolutionary theory" (1992: 210). Such a

---

11. In many respects, Mackinder 1904 can be read as a land-based reworking of the more subtle writings of A. T. Mahan on the role of sea routes in military history.

synthetic ambition marks a broad tradition within social evolutionary accounts of early complex polities.

In place of Mackinder's geographical determinism, social evolutionary approaches within the organic tradition mustered the much more robust concept of adaptation to mediate between the space of the environment and the space of human social worlds. An early expression of the environmentalist view arises in V. Gordon Childe's account of the evolution of early complex polities. Childe introduced to the study of social transformations the idea that spatial variation, if described in environmental terms, might have some impact on the consequent evolutionary development of a society (noted in Service 1975: xvi). In particular, Childe points to the role played by the uneven distribution of natural resources in promoting or retarding historical progress through the stages of productive revolution.[12] For example, he argues (1931) that Scotland would never have experienced a Neolithic revolution built on food production without considerable diffusion of knowledge and materials from the south because the land lacked domesticable plants and animals. Childe's sense of the link between environment and social space is commonsensical—it neither relies on a universal account of process nor preserves the unity and necessity of social evolution, because much is left to hang on highly contingent historical moments, such as the sharing of knowledge and plants between differentiated social groups.

A similar effort to situate human-environment relations within historical research coalesced in France in the mid-twentieth century around the journal *Annales: Economies, Sociétes, Civilisation*. For Fernand Braudel, the most prominent member of the *Annales* school, the problem of history did not lie in narratives of individual actors and particular events but rather in "landscape," which he defined as the impersonal forces that fashion human existence. Historiography, he continued, must plot these slow temporal rhythms that lie beneath the immediate activities of microhistory in the times of *conjunctures* and *longue durée*.[13] It is only in histories of the longue durée where a recognizable sense of landscape emerges for Braudel. The quicker pace of conjunctures allows us to track demographic trends and shifting patterns of trade. However, in Braudel's most ambitious study (1972–73), spatiality enters into these discussions only as

---

12. A point to be reintroduced to critical Marxism almost half a century later (e.g., Smith 1989).

13. Braudel's sense of landscape derived most directly from the human geography of Paul Vidal de la Blache (Clark 1985: 181).

the ecological, physiological, hydrological, and climatic parameters of the environment constrain economic and social possibilities. As he put it: "[W]hen I think of the individual, I am always inclined to see him imprisoned within a destiny in which he himself has little hand, fixed in a landscape in which the infinite perspectives of the long term stretch into the distance both behind him and before. In historical analysis as I see it, rightly or wrongly, the long run wins in the end" (Braudel 1972–73: 1244). As has often been remarked (see, for example, Hexter 1972: 518), the heroes of Braudel's history are mountains, plains, peninsulas, and the Mediterranean Sea itself; these features are given purpose and agency in human affairs, obstructing action and limiting possibility by imprisoning humans in the vast temporal expanse of the longue durée.

The influence of Braudel and the *Annales* school on the study of early complex polities has been significant if largely indirect and regretfully delayed. To date, the most intensive effort to transport *Annales* historiography into antiquity is Peregrin Horden and Nicholas Purcell's (2000) monumental prequel to Braudel's study of the Mediterranean world ca. A.D. 1550. Horden and Purcell offer an extraordinarily erudite, broadly encompassing, historical ecology of the region that explores the impact of local environments on antique and medieval regional history. In many ways, they retread the same analytical ground as Braudel, utilizing more sophisticated measures of environment to argue for a deeper temporal sense of historical continuity. Beyond the confines of Mediterranean history, recent isolated efforts to formalize the contribution of *Annales* historiography to archaeological research have focused primarily on Braudel's typology of temporality, co-opting the longue durée to the cause of process and persistence over recent challenges by advocates of agency and contingency (Bintliff 1991; Knapp 1992). Insofar as these efforts implicitly embrace Braudel's sense of landscape rooted in physical geography rather than in social worlds, they have revived an organic absolutism that was systematically drawn into the study of early complex polities by Julian Steward.

Steward formalized Childe's account of the interaction of environment and evolution, advancing an ecological reworking of social transformations that argued for multiple lines of historical development differentiated by adaptation to local ecological conditions. Steward argued that the "core" of a culture centers in "the constellation of features which are most closely related to subsistence activities and economic arrangements" (1972: 37). As a result, cultures arising in distinct environmental settings, such as rain forests, temperate forests, or deserts, would develop along different evolutionary pathways, or trajectories. The trajectory of general evolution, therefore, could not be singular, because in each ecological zone the functional relationship between key variables would create social evolutionary lines unique to that environment—but generalizable across similar environmental conditions (ibid., 208–9). Steward elaborated this conception in reference to complex societies, arguing that pristine civilizations could only have arisen in arid regions, where, following Karl Wittfogel (1957), he argued that the demands of irrigation provided the managerial foundation for the State (see also Steward 1972: 23–24, 203–5).

It is important to note both the similarities and the differences between the ecological position of Steward and the *Annales* school and the mechanical absolutism advanced by locational approaches. Where the mechanists see variation rooted in local environmental conditions as noise that needs to be filtered out in order to reveal the fundamental geometric laws of space, historical ecologists treated local conditions such as topography, climate, and hydrology as the basis on which variation in spatial forms—and historical variation in evolutionary process—could be understood. Where the mechanists rely on behavioral assumptions to translate spatial law into geometric form, organicists employ the much more flexible concept of adaptation to define relations between form and nature. The rather profound implication of this re-theorization is that, rather than regular laws shaping the space of the social world, the social world fits itself to the space of the environment.

The intellectual tradition that followed Steward was less a revitalization of pure environmental determinism (although such positions are not unknown), than the now almost canonical assumption that a broad set of interrelated spatial parameters rooted in the material interactions of humans and environments—resource distribution, demography, hydrology—are the prime explanatory variables in accounting for the rise of early complex polities. We can trace the impact of Steward's ecological account of social evolution in a number of sources. During the 1950s and 1960s, both William Sanders (1956, working from a Mesoamerican viewpoint) and Robert McC. Adams (1960, 1966: 51, operating primarily from a Mesopotamian perspective) embraced an organic spatial ontology, suggesting that explanations of the rise of complex societies must look to the natural world and organization of subsistence relations that mediate between society and environment. (See chapter 5 on the development of Adams's more recent investigations.)

Among the broad field of more current social evolutionary explanations of archaic State formation, we can outline a range of positions derived from an organic absolutism. What unites these approaches is a shared conviction that the material circumstances of human-environment interaction—as expressed primarily in economy, ecology, demography, and technology—drive social evolution and thus determine the spatial dimensions of social life. It is as a result of this privilege accorded the internal dynamics of human-material relations that organic absolutist ontologies of space are most commonly expressed within a set of explanations for the archaic State that focus on conflicts stemming from crises of resource availability. For example, population pressure, a spatial concept constituted in the numbers of people within a geographically delimited resource area, has been the most popularly cited "prime mover" in explanations of the rise of the archaic State, operationalized within a wide variety of theoretical configurations. General theories of the emergence of States that point to population pressure as the lead factor generally see demographic growth as creating material scarcity (of land, of food) that is resolved through some component of State organization. For Morton Fried (1967: 196), and later Don Dumond (1972), the result of population pressure is social stratification, a substrate of inequality within which the formal elements of political complexity may emerge. For Robert Santley (1980), as well as for Allen Johnson and Timothy Earle (1987), the result of population pressure is an intensification of agricultural production and increased competition—a foundation for subsequent expansion of bureaucracy and governmental controls. For Robert Carneiro (1970) and, more recently, Patrick Kirch (1988), the result of land shortages caused by barriers to migration in circumscribed environments is warfare, which in turn produces stratification (in the form of conquered and conquering social groups) and occupational specialization.

These demographic theories of social evolution share a similar organic account of space—the natural world defines the parameters of sustainable geographic relationships between people and environments determinative of social evolutionary process. When the parameters of this spatial relationship are breached—that is, more humans in a region than productive capacities can support—the result is profound sociohistorical change: "[P]opulation growth and a chain reaction of economic and social changes underlie cultural evolution" (Johnson and Earle 1987: 5). Yet there is a broad theoretical spectrum between, on the one hand, the direct appeals to population pressure by Fried and approaches such as Carneiro's that attempt to set such pressures within the parameters of a particular spatial configuration (such as a circumscribed environment). This axis of variability oscillates around the fundamental inconsistency in attempting to yield environment a determinative position in social process while refusing to concede the spatial variability underlying environmental difference any causal role. Such ambivalence regarding the theoretical significance of space, broadly conceived to incorporate both geography and environment, lies at the heart of the organic ontology as it has been mustered to the cause of social evolution. Consider Johnson and Earle's discussion of the evolution of the archaic State:

> Our two examples [of the archaic state], medieval France and Japan, are widely separated both spatially and culturally. Yet when the layers of aesthetic, technological, social, and philosophical differences are stripped away—when all that remains is the small set of variables that form the core of our model of social evolution—we find astonishing similarities between the two societies. . . . [Both] developed gradually into states, propelled by pressures and opportunities arising from population increase and intensification of land use. (Ibid., 248, 256)

It does not seem particularly astonishing, on an epistemological level at least, that once everything different is "stripped away," two social formations might look quite similar. Rather, what is particularly intriguing in the argument forwarded by Johnson and Earle is the very willingness to strip away spatial difference in terms of geography (thus maintaining the social evolutionary claim to universality) even as the spatiality of the productive environment, specifically the enlargement of land under irrigation agriculture and the expansion of secure exchange networks that facilitated large increases in population, is ceded general determinative status in the rise of the archaic State.

What we see in these organicist frameworks is the replacement of the neo-classical models of human behavior that mediated ties between evolution and spatial form in mechanical models with an ecological account of social formation that roots dimensions of the human landscape in the material circumstances of production. Market principles yield to ecological rules regulating reproduction and subsistence. Although the basic Newtonian theory of space as an a priori object unchangeable in its fundamental nature remains intact, space, when read as environment, is accorded a circumscribed position in explanations of social transformations. The organic understanding of absolute space is considerably more flexible than the mechanist tradition, even as it is less epistemologically consistent. However, though the parameters of human agency in the human-environment equation, narrow in early geographic determinism, were widened by Lucian Febvre's possibilism (1925: 236–37) and Claudio Vita-Finzi's opportunism (1978: 11), responsibility for the rise of early complex polities invariably is ceded to an extra-social condition, a universal drive rooted in environmental transformation. Most organicist conceptions of the relation between the natural space of the environment and the social space of the human world do, as Stephen Sanderson (1995: 12–13) protests, describe some interplay between these components. But he protests too much, as determination in the last instance must always be rooted in rules that order the natural world as opposed to a practical social logic. The effect is to render individuals and groups as passive respondents to change—as dull witnesses to history rather than makers of it—able to develop only limited responses to ecological, demographic, or technological problems. Yet even after several decades of organicist accounts of early complex polities, Childe's observation that "men seem to be impelled to far more strenuous and sustained action by the idea of [a] two-headed eagle, immortality, or freedom than by the most succulent bananas!" (1946: 8) remains a powerful, unanswered, commonsense rejoinder.

A CRITIQUE OF ABSOLUTE SPACE

Mechanical and organic positions share a basic sense of the unity of space. But, though the abstract geometry of the former removes space from any influence on social evolution, the functional determinacy of the latter allows a restricted set of spatial variables to play a role in shaping historical transformations. Although it has generated a number of extremely important insights into the connections between spatial form and socio-historical change, a number of objections to the absolutist position point toward its philosophical instability (Werlen 1993: 2; *pace* Nerlich 1994). In addition to the specific criticisms of its mechanical and organic manifestations discussed above, three general philosophical critiques of spatial absolutism should be noted. The first questions space's objective independence. If space exists as an independent object of research, then we should be able to point out its location in the physical world. But this is impossible without reference to the objects that inhabit space. For example, to argue that space is an object is to assert that, because Chicago is north of Tucson, there must be in the world not only places and their relations but a real entity we can term "northernliness" that is separate and distinct from either city. Yet such an object is not locatable outside of the spatial arrangement of other objects (see Sklar 1974: 167).

A second critique is epistemological, questioning the ability of the researcher to access a dimension of space that is somehow more fundamental than the immediate spatial relationships that are the objects of empiricism. Henri Lefebvre dissented from attempts to impose the epistemology of science on examinations of space: "To date, work in this area has produced either mere descriptions which never achieve analytical, much less theoretical status, or else fragments and cross-sections of space. . . . [T]hough [these techniques] may well supply inventories of what exists *in* space, or even generate a *discourse on* space, [they] cannot ever give rise to a *knowledge of* space" (1991: 7, emphasis in original). In other words, though absolutist approaches to the spaces of human social life may be able to describe what spaces look like to varying degrees of abstraction, they cannot provide an account of why spaces arise in certain forms, how places form within the cultural milieu, or what role they play in social life. This would seem a particularly profound problem for an archaeology committed to exploring what ancient polities did to establish and reproduce their authority.

Third, the absolutist position provides an analytical framework for examining only physical spatial form. But the physicality of space—the concreteness of the created environment—is only one dimension of landscape. As the burgeoning attention to the interaction of space, imagination, place, and memory demonstrates—a tradition that extends through Edward Said's *Orientalism* and Raymond Williams's *The Country and the City* to Simon Schama's *Landscape and Memory*—the physical form of space is only one dimension of the much broader landscape of human social worlds. Where, within either the mechanical or organic positions, can we articulate form in a historically or socially meaningful way with the values and beliefs that are tightly bound into them or the representations that carry spatial form into media? Such representations of spaces are invariably stripped away within absolutist approaches as epiphenomena of the progress of social evolution. Examinations of the landscape as multifaceted places—vested with cultural significance, social memory, and political consequence—cast substantial doubt on the ability of absolutist accounts to understand the social.

This critique of absolute space holds profound implications for social evolutionary theory. However, an encompassing critique of social evolutionary perspectives in studies of early complex polities is not germane to the present discussion and thus must be left for another forum. Yet the problems that we can discern in the absolute space of evolution suggest we might also profit from a closer examination of a second recently reenergized tradition in studies of early complex polities: historicism.

Historicism and Subjective Space

Although social evolutionary frameworks have come to dominate accounts of early complex polities since the 1950s and 1960s, the field emerged during the nineteenth and early twentieth centuries as a resolutely historicist domain of inquiry, built on a subjectivist (or substantivist) ontology of space.[14] By historicism I mean those philosophies of history that define all social and cultural phenomena as historically (not universally) determined and thus understandable only in terms of their own time and place (Stanford 1998: 155).[15] The historicist tradition arose in opposition to attempts to extend the principles of natural scientific explanation to human affairs as exemplified in social evolutionism. Historicism's hostility to the social evolutionary account of history was rooted in the latter's focus on mechanical accounts of causation at the expense of meaningful descriptions of human affairs. The sociohistorical significance of events lies, according to the historicist, not in the interaction of human groups with an exterior domain of natural laws (physical or environmental) but in the significance actions hold for individuals: "When an historian asks 'Why did Brutus stab Caesar?' he means 'What did Brutus think, which made him decide to stab Caesar?' The cause of the event for him means the thought in the mind of the person by whose agency the event came about: and this is not something other than the event, it is the inside of the event itself" (Collingwood 1994: 214–15). R. G. Collingwood perceptively argued that to describe the murder of Caesar (or more generally the transition from Republican to Imperial Rome) as epiphenomenal to changing environmental conditions around the eternal city or inherent rules of political formation evacuates from history all that is interesting and illuminating. But what consequences does this conclusion hold for understanding the spatiality of social life?

The roots of historicism are traditionally traced to the philosophical writings of Giambattista Vico, who argued that all aspects of every society are characterized by distinct patterns and styles particular to it; each successive stage of social history grows from its predecessors by human

14. This is particularly the case for studies of early complex societies in the Old World (see Barkan 1999; Kuklick 1996; Marchand 1996).

15. Note that the sense of historicism used here is very different from Karl Popper's (1957: 3) definition. Popper rejected the search for regular laws of history—of social evolution. Historicism as it has been generally developed in the twentieth century concurs with Popper, emphasizing the particularity of all societies in space and time and refusing to subsume them into a general account of historical change.

agency, not natural causation. The cultural creations of humans, including spatial forms, are thus, above all, forms of expression (Vico 1996; Berlin 1976: xvi–xix). Although the search for evolutionary laws underlying social phenomena demands the erasure of spatial difference, historicism describes spatial forms as presentations of fundamental dimensions of human beliefs and values.

A subjectivist spatial ontology was originally set out by Immanuel Kant, who argued, contra Newton, that we do not experience objects in themselves but only the impressions they occasion in our senses. Kant concluded that the connections we perceive between these sensory representations do not inhere in the world itself as an objective universal container but rather are facets of an objective order imposed by the observing subject. Thus space is an element of the subject's apparatus of perception and prior to any objects themselves: "Space is not an object of outer sensation; it is rather a fundamental concept which . . . makes possible all such outer sensation" (Kant 1992: 371). Space is thus redefined from the Newtonian sense of a privileged object to become a subjective dimension of representation, a form of intuition that "reflects the nature of the knowing subject rather [than] the object known" (Peet 1998: 18). Although Kant clearly locates this spatial intuition in the cognitive apparatus of individual subjects, it became dislocated as it was folded into historicism by Johann Herder.

Herder, who attended Kant's lectures on the role of geography in human history, removed spatial intuition from individuals and bestowed it as a normative "spirit" shared by a particular people in a particular age.[16] Herder argued that knowledge of the past does not arise in reference to a universal theory of human nature or the course of history but rather must be grounded in localized accounts of qualitative variation in subjective histories. As described by Schama, Herder's contribution to historicism was his vision of "a culture organically rooted in the topography, customs, and communities of the local native tradition" (Schama 1996: 102–3). Herder located the authentic German spirit in a vision of a medieval landscape dominated by vast, unspoiled forests. The primary unit of historical analysis and interpretation was thus constituted by the com-

16. Archaeological theorists often, though errantly, describe historicism as historical particularism due to its resistance to overarching accounts of world social change that neglect the particulars of each case. Historicism is also predicated on broad meta-narratives, but these are localized within discrete, presumably enduring, cultural groups rather than broadly generalized into global histories.

munity of shared beliefs, values, and experiences—what Herder (1966) termed nation (see also Berlin 1976: 182–83; Stanford 1998: 157). As ultimately elaborated by Hegel, historical process is defined as the working out of this nationally rooted "spirit," a presentation of the internal life of a generalized set of shared beliefs and values.

Within the historicist tradition, we can describe two distinct strands at work in studies of early complex polities: a romantic historicism that arose out of antiquarianism and biblical hermeneutics, and a revivalist neo-historicism that draws heavily on phenomenological and semiotic theory.

ROMANTIC SUBJECTIVISM

As fascination with antiquity blossomed in the nineteenth century, historicism provided the primary philosophical foundation for investigations of early complex societies. Much of the original impetus to explore ancient civilizations, particularly those of Egypt and the Near East, arose from biblical hermeneutics, which saw in archaeology and epigraphy methods for demonstrating the veracity of historical accounts in the Bible (Kuklick 1996). The links between the study of ancient civilizations and theology were manifest not only in the large number of religious officials who formed the early ranks of epigraphers but also in the public attention garnered by biblically related discoveries, such as the Babylonian flood myth recorded on a clay tablet from the site of Nineveh (published in 1872 by George Smith). Biblical interpretation continued to play an orienting role in studies of ancient civilizations well into the twentieth century. For example, in 1928 C. Leonard Woolley interpreted a thick silt layer at the city of Ur in southern Mesopotamia as a result of the flood described in the Bible (Woolley 1965: 130–34).[17]

The larger corpus of Woolley's writings, along with the works of such pioneers in archaeological and epigraphic research as Claudius Rich, Paul Emile Botta, Robert Koldewey, William Flinders Petrie, and Austen Henry Layard, do not reveal an overarching concern with biblical interpretation in light of recovered materials. The orienting questions of their research were neither biblical nor metahistorical but rather culture-historical: who built a given settlement or monument and when? As a result, the majority of their writings are primarily descriptive in character,

---

17. More recently, geological investigations into Holocene climatic cycles in the Black Sea basin have been explicitly phrased as providing evidence of Noah's flood, a sensational if historically dubious contention (Ryan and Pitman 1998).

recording what was found, where, and what it looked like. But where description yielded to more interpretive remarks, most of these writers express a historicist fascination with the expressive artifacts of now-vanished civilizations that was rooted in an understanding of a collective spirit or genius. Essential differences in this spirit, particularly between Sumerian and Assyrian manifestations, were, for Woolley, most succinctly expressed in the contrasting images of landscape. In his discussion of the Ur-Namma Stela, a stone carved in low relief to commemorate the exploits of King Ur-Namma of the Third Dynasty of Ur (ca. 2100 B.C.; see chapter 5), Woolley drew a stark contrast between the scenes on this monument that depict the Sumerian king as a sacred builder and Assyrian images of cities sacked and rivers choked with bodies: "[T]o turn from [the Stela of Ur-Namma] to the wall-reliefs of [the Assyrian king] Ashur-natsir-pal . . . is to understand at a glance the difference between the Sumerian and Assyrian character" (1965: 135–36).[18] In this construction, the aesthetics of landscape, as carried in dominant forms of representation, directly and unproblematically reflect essential national or cultural differences.

Within the early historicist tradition, architecture and spatial form are regularly interpreted as expressions of the unique genius of a people, culture, social group, or civilization. For example:

> The planning of Mesopotamian Temples takes an even more definitive form in the so-called Protoliterate period, which dates from the final centuries of the fourth millennium B.C. This was a time when *the genius of the Sumerians* seems to have reached its zenith, finding expression in some of the cardinal inventions that have contributed to our own civilization. (Lloyd 1980: 12–13, emphasis added)

> Given the high intelligence and religious fervor of the ancient Maya, it was almost inevitable they should develop a great religious architecture. (Morley 1946: 315)

The romantic subjectivist ontology of built space can also be seen in a number of studies of ancient urbanism. In his discussion of the Greek *polis,* H. D. F. Kitto points to distinct characteristics of the Hellenes to

---

18. Woolley continues later in the book: "Sumerian genius evolved a civilization which persisted for nearly fifteen hundred years after its authors had vanished" (1965: 189). This heritage lingers today, according to Woolley, in such architectural features as the arch, the dome, and the vault—formal and aesthetic innovations derived from Sumerian models expressive of both a native Sumerian genius and that civilization's contribution to humanity.

explain the dense patterns of settlement there in contrast to other regions: "At this point we may invoke the very sociable habits of the Greeks, ancient or modern. The English farmer likes to build his house on his land, and to come into town when he has to. . . . The Greek prefers to live in the town or village, to walk out to his work, and to spend his rather ampler leisure talking in the town or village square" (1951: 68). Formal differences between Greek- and British-built environments are explicated first and foremost to fundamental differences in the character of each nationality. Of course, such aesthetic evaluations can cut both ways, allowing for both praise of artistry and condemnations of unsophisticated society and culture—a point that Mario Liverani has raised in regard to aesthetic characterizations of the Assyrians: "In discussing the oppressive burden of the despotic state on culture and society, [Jacob] Burckhardt mentions 'the rude royal fortress of Nineveh' and 'their miserable architectural structure and their servile sculpture'" (1997: 86).

A number of difficulties render traditional romantic subjectivism indefensible as a general framework for spatial theory. First, it relies on a direct relationship between a people and place, treating the relationship between a geographic locale (such as Greece or Sumer) as unproblematically coterminous with a normative set of beliefs and values. This spatial essentialism is an extraordinarily difficult position to defend because relations between people, place, and culture are increasingly understood as social and political productions. The process of building what Benedict Anderson (1983) has described as "imagined communities" can be traced deep into the record of early complex polities. For example, after the sack of Babylon by the Hittites in 1595 B.C., the city became the religious capital of a new kingdom under the control of a distinctly non-Mesopotamian group known as the Kassites. Far from using their political dominance to express a distinctly "Kassite" spirit through their architecture, they went to great lengths to perpetuate Sumerian traditions of building and adopted the Akkadian language. At their administrative center, Dur-Kurigalzu, the Kassites built a ziggurat—a quintessentially "Sumerian" architectural form. The production of traditional Sumerian spaces was clearly not a stable expression of the Kassite's "essential character" (Oates 1986: 86–104).

A second, more troubling criticism of romantic subjectivism is that it overly aestheticizes form. By attending only to those details considered culturally expressive, buildings and monuments are no longer understood as settings for activities and actions. Romantic subjectivism thus privileges perception—the affective qualities of sublimity—over experience and imagination. One consequence of this position is that analysis is restricted entirely to monumental architecture and urban organization; domestic architecture and settlement patterns apparently lack the grand expressivity of temples, palaces, and cities. A further consequence is that, in linking the aesthetics of spatial form to national character, romantic subjectivism sets the stage for evaluative designations of superiority and inferiority. Thus the flip side of the Sumerian genius visible in beautiful architecture is the conclusion that an unappealing built environment bespeaks a degenerate character. It is this aesthetic moralism, and its unfortunate political uses, that Suzanne Marchand (1996) captures in such revealing detail in her study of nineteenth-century German archaeology and its foundational philhellenism.

Perhaps the most profound critique of romantic subjectivism for the purposes of this study is its deeply embedded tendency to naturalize contemporary politics under the rubric of empathetic understanding. This is an objection that Walter Benjamin (1985) raised in his "Theses on the Philosophy of History." To follow the romantic historicist injunction to relive bygone eras is, in Benjamin's view, to collapse into a profound "indolence of the heart . . . which despairs of grasping and holding the genuine historical image as it flares up briefly" (vii). For Benjamin, the object of empathetic history could only be the victor. By sustaining empathy for victors in the past, historicism can serve only as a tool of the powerful, transforming the artifacts of violent political struggle and oppression into fetish objects, into cultural treasures (xvi).

The historicist fascination with the past is self-consciously humanist in its expression, in contrast to the scientific models that came to dominate prehistoric archaeology in the latter half of the twentieth century. It is thus not surprising that romantic subjectivism, though still a profound undercurrent within more antiquarian sensibilities, was largely eclipsed by the social evolutionary turn in the study of early complex polities in the Soviet Union as early as the 1930s and in the Anglophone West during the 1960s.[19] Yet just as investigations of social complexity in the ancient world were giving themselves over to evolutionism and spatial absolutism, withering critiques of the ontologies of modernism—of its temporocentrism, its fetishism of the systemic, and its stultifying metahistorical discourse—encouraged a revival of subjectivist perspectives in social and, more recently, archaeological theory.

19. One of the earlier Soviet works to adopt an avowedly social evolutionary outlook was a study of the East European steppe (Kruglov and Podgayetsky 1935; see also Binford 1965).

NEO-SUBJECTIVISM

Within a steadily growing corpus of neo-subjectivist investigations of space, two strands of thought have exerted the most influence on the study of early complex polities: a communicative tradition which argues that spatial forms arise as forms of nonverbal expression, and a phenomenological approach that interprets created environments as expressions of cultural systems of belief or cosmology.

Communicative Traditions.   The communicative tradition of spatial analysis has grown in numerous directions, ranging from the spatial syntax approach of Bill Hillier and Julienne Hanson (1984, which argues for a universal human grammar of spatial units capable of generating the entire repertoire of spatial form) to the nonverbal communication approach set forth by Amos Rapoport (1982). What draws this tradition of spatial analysis together is a shared emphasis on space as a means of transmission parallel to language. That is, space not only passively expresses a certain aesthetic but also communicates information about itself and the social world within which it is embedded. Thus users—those sufficiently acculturated to understand the semiotic system—know to react in certain culturally proscribed ways within certain built environments. Space, in this sense, is a medium that rests on shared cognitive faculties for encoding and decoding what Darryl Hattenhauer (1984) refers to as "the rhetoric of architecture."

The contention that space not only expresses but also argues is an important expansion of the subjectivist ontology, one that allows a less idealist sense of the engagement between form and aesthetics to emerge. An important mediator in the appropriation of communicative subjectivist positions for the study of early complex polities was Henry Glassie's (1975) investigation of folk housing in Virginia. Based on an examination of over 100 eighteenth-century homes in the Virginia tidal region, Glassie described a restricted set of spatial rules—a grammar—that he contended expressed the worldview of contemporary local society. Contained within the precise architectural rules that governed the placement of windows, the size of doors, and the axial dimensions of rooms, Glassie found a mental structure grounded in a shared architectural repertoire. Although his analysis has produced a flurry of archaeological investigations within Americanist historical archaeology, it—and communicative subjectivism generally—has had a more muted impact on the study of early complex polities (see also Leone 1988: 235–37). Although a number of studies have deployed the techniques generated by semiotically rooted analyses, such as Hillier and Hanson's permeability diagrams (see chapter 6), most have either explicitly jettisoned the attached neo-historicist ontology or attempted to re-cast the communicative approach as a relational theory of the social (discussed below) rather than a subjectivist account of inherent meaning. (Examples of the former are Smith 1996, 1999. Examples of the latter are Ferguson 1996; Van Dyke 1999.)

An alternative strand of the communicative approach argues that built environments are best understood as texts and so should be "read" as such (see Blier 1987; Cosgrove and Daniels 1988; Darnton 1984; Meinig and Jackson 1979). Like other variants within the semiotic tradition of spatial analysis, the textualist position regards spatial form and aesthetics as encoded information whose understanding is shaped in reference to larger discursive fields. Thus royal monuments are "readable" in reference to the larger cultural discourse on rulership, a connection made intelligible through an enduring series of repeated tropic conventions. Although echoing the textual turn advanced by early post-processualist approaches to archaeological interpretation, the metaphor of landscape as text has not greatly influenced examinations of early complex polities.[20]

Phenomenological Traditions.   A second strain of neo-subjectivism that has percolated more forcefully into archaeology arises from more phenomenological sources and suggests that spatial forms can be understood as representations of systems of thought, belief, or worldviews. (On the body as a spatial model, see Bloomer and Moore 1977; Hall 1966; Tuan 1977.) The theoretical roots of this position lie most profoundly in Martin Heidegger's emphasis on the affective ties between people and place and the role played by spatial phenomena in organizing meaning. Phenomenological approaches to space are unified by a concern to approach built forms as representations of lived experience. (Compare related hermeneutic discussions of architecture, such as in Gadamer 1975: 138–40.) Space, they argue, is not defined by the geometry of form; rather, it is a sensual experience in which perception resonates with cultural values. Where semiotic approaches decode space as a series of signs, phenomenologists claim to go beyond the sign to reveal a richer account of life in the world. This

20. However, James Duncan's (1990) examination of the landscapes forged by the Kandyan kingdom in early-nineteenth-century Sri Lanka does provide a detailed application of the approach's central premises. See also the brief discussion of Duncan's textualist approach in J. C. Chapman's (1997) study of ancient landscapes in Eastern Europe.

account rests in the construction of links between form and imagination as mediated by sense experience. Thus the "odour of raisins drying in a wicker basket" leads Gaston Bachelard (1969: 13) to elaborate the dreams that define our experiences of house and home.[21] Although phenomenology has become an increasingly prominent part of archaeological thought, to my knowledge it has yet to be formally deployed in investigations of early complex polities. But its recent surge in popularity as a framework for examining ancient landscapes demands close attention.

The first strand of phenomenological neo-subjectivism examines built environments as cosmological models, expressive of culturally specific understandings of the order of the universe. In one of the more simplistic analyses of such spatial expressions of worldview, Vincent Scully (1991: 26) has suggested that differences in early Mesopotamian and Egyptian monumental architecture were rooted in fundamentally different cosmologies that imprinted themselves directly on their architecture. However, more nuanced accounts of the expression of cosmology in built space can be found in a number of studies of Classic Maya site planning and settlement patterns. The twin-pyramid groups at Tikal have been interpreted as depictions of the multilayered Maya cosmos turned on its side (Ashmore 1989: 272; Coggins 1980; Guillemin 1968). Evan Vogt has attributed these spatial iterations of cosmological models to a handful of deeply rooted structural principles: "[Mayanists] are currently on the track of a cluster of structural and conceptual principles revolving around settlement patterns and their concomitants in social, political, and ceremonial life and cosmology that can explain much of the Maya past and present" (1983: 114). Vogt's analysis articulates, in classic subjectivist fashion, a set of built forms with a culturally specific account of mind. Although phrased now in structuralist terms, such analyses are not ontologically different from earlier analyses that invoked the genius of the Maya.

Perhaps the most intricate interpretation of architecture as a model of the cosmos is Eleanor Mannikka's account of the Khmer site of Angkor Wat, built in the twelfth century A.D. Mannikka discovers in the buildings at Angkor a structuring numerology that centers on number pairs derived from observations of celestial cycles. In this way, Mannikka argues, the builders of Angkor transformed time—the movement of celestial bodies—into built space and provided an architectural model of the universe: "Angkor Wat is also an architectural expression of the universe

21. This incident closely resembles the madeleine that transports the author in Marcel Proust's *Remembrance of Things Past*.

created by Viṣṇu: the celestial epic, the lunar and solar constellations, the cycles of time, the realm of King Sjryavarman, and the historical dates that surround the tower are a reflection of the celestial realm created by Viṣṇu. The architects wanted to join that celestial world to Angkor Wat, not just through measurement and reliefs, but through architectural design as well" (1996: 264). The numerology that Mannikka finds in the geometry of Angkor Wat makes an excellent case for articulating dimensions of spatial form with cultural structures of religion and belief. However, as in most phenomenological approaches, there is little discussion as to why it was politically important or socially desirable to build such intricate connections into the landscape.

The most highly developed spatial phenomenology has been forwarded not in reference to early complex polities but rather in the context of Old World Neolithic societies. Christopher Tilley's *The Phenomenology of Landscape* (1994) is the most highly theorized of a number of works on the subject and so warrants discussion here. Tilley's central argument is that, in Europe, the transition from Mesolithic to Neolithic communities was experienced as a significant change in the relationship between landscape and people. In particular, new forms of burials and the advent of monument building marked an appropriation of ancestral powers that were "sedimented" in the topography and symbolic geography of the land. Tilley's analysis tries to draw out these sedimented meanings and thus recreate the mythology that underlay the creation of landscape form.

Although Tilley's determination to break free from prevailing environmental determinisms should be applauded, the reduction of landscape to mythopoeia is ultimately unconvincing. Because he draws his analysis rather directly from land into myth, we lose any sense of the social milieu. That is, by rooting landscape solely in affect, Tilley is forced to rely on a rather direct and uncomplicated account of the emotional connections between human and land. Thus emotional attachment to ancestors, for example, takes on the status of a deep-seated component of an essential psychology rather than a socially constituted value created in Mesolithic cultural practice. The primary difficulty with the phenomenological approach to landscape is that it assumes stability either in the affective responses of gazing subjects—so that certain dimensions, such as height, will always cue similar embodied responses—or in the environment itself—such that certain environments come pre-loaded with specific cultural meanings.

The drive to root place in a stable subject emerges in the tendency to treat the body as the irreducible measure of landscape. Hence Tilley reconstructs the links between several features of the Neolithic Dorset Cursus as requiring a procession from northeast to southwest because key features of the route "would have little or no visual or somatic impact when approached from the NW [northwest] and would not surprise" (1994: 197). When projected deep into the past, the phenomenological method rests heavily on assumptions of enduring links between sets of emotions and spatial forms.[22] But what seems to be forgotten here is that the body itself bears the marks of social process. Bodies are split, divided, totalized, and broken down within specific sociopolitical formations (Foucault 1978, 1979a). Hence the body cannot provide a universal foundation for landscape interpretation because it is highly dependent on particular understandings of subjectivity and of social processes of subjectivization. Tilley's analysis of the Dorset Cursus assumes, if not a universal mind, then at least a universal aesthetic of ritualized bodies. Although these spatial values may indeed ring true, as social products they not only are culturally expressive but also do social work—they reproduce inequalities, glorify and legitimate rulers, and create subjects.

The tendency to assume that landscapes exist pre-loaded with meanings emerges most emblematically in Tilley's account of Mesolithic communities in Pembrokeshire: "I want to suggest that it was precisely because the coast provided both rich economic resources and a wealth of named and distinctive natural topographic markers that it was so symbolically important to both Mesolithic and Neolithic populations" (1994: 86). By focusing on meanings sedimented in the landscape, Tilley's phenomenological approach obscures the process by which spatial meanings are produced, reinforced, and manipulated. As a result, the meaning of the landscape always seems to emerge prior to the symbol-making activity of actual people. Although it is clear how specific groups are attached to landscape, it is manifestly unclear how they came to be connected in the particular way that they did. Thus, we are left without a robust interpretation of the particularity of form, only a defense of its affective potency (which, of course, was assumed at the outset).

The phenomenology of landscape leaves us with a rather apolitical vision of landscape where the sedimentation of meaning in place tends to preclude an account of the interests that shape the particular contours of landscapes. Indeed, power is rather absent from most neo-subjectivist

22. For Tilley, visibility is an enduring tectonic value in the Mesolithic and Neolithic eras, but it is unclear exactly why.

approaches to space in general. When power does arise in these analyses, it does so primarily in reference to modern manipulations of ancient places, such as the deployment of Stonehenge in a range of political agendas, rather than in reference to politics in the ancient world (see, for example, Bender 1998). Hence we moderns are often accused by the neo-historicists of having evacuated meaning from the landscape—a repeated, if nonsensical, charge—leaving only strategic encounters with place where once there were enduring sets of traditional histories, myths, and values. Although the reactionary tones of this romantic desire for a return to a more meaningful place should not be overlooked, the central point of this critique is the difficulty in accounting for the actual production of landscape within neo-subjectivist ontologies of space.

Perhaps the most important steps toward an account of landscape within archaeology have been taken by Richard Bradley (1998, 2000) in his two complementary volumes on monuments and natural places. In his analysis of the European Mesolithic-Neolithic transition, Bradley is primarily interested in examining the socially organized relationships between people and place. Hence, monuments erected by Neolithic communities are not simply marking locales of transcendent meaning; they are built elements of social practice intended to "dominate the landscape" and assert control (Bradley 1998: 34). What is particularly noteworthy about Bradley's approach to landscape is that created environments and natural places are described as active elements of social production. That is, they do not arrive pre-packed with enduring cognitive patterns of structural signification[23] but rather take on symbolic content within practical relationships—"the relationship between the form taken by the architecture and the kind of audience to whom it was addressed" (ibid., 101). At times, Bradley's move away from a subjectivist ontology of space seems incomplete when motives to social actions, such as "dominating the landscape," float unmoored from any account of why the landscape should be thought of as a potential subject in itself rather than a complex medium for subjectivization. Indeed, what is most profoundly missing is a fully realized discussion of the social interests underlying programs of landscape production. But Bradley's immensely readable analyses effectively point in the direction of a revised archaeological account of space, one that moves away from a dependence on either an absolute space or a prior cognized place toward a set of spatial practices constituted within social worlds.

23. See also Hodder 1990 for an interpretation of the Neolithic *domus*.

A CRITIQUE OF SUBJECTIVE SPACE

The neo-subjectivist turn in archaeological spatial analyses has offered a number of important insights that should be preserved. First, these approaches have irrevocably problematized absolutism's refusal to engage with space in any dimension other than the purely material. The neo-historicist perspective clearly indicates how space must also be understood as emerging as a perceptual category, one that not only directs and organizes the flow of bodies but also is intelligible. That is, space interacts with our senses to provoke affective reactions, to stimulate feelings and associations, and to substantiate histories. This broadening of the sense of space is a vital contribution. Second, the neo-historicists have brought a revitalized sense of the links between cultural and formal difference, providing accounts of how different groups of people construct very different ways of living in the world. However, despite these contributions, neo-subjectivist frameworks tend to be undone along a number of lines.

The most fundamental problem with neo-subjectivism is that these approaches provide little account of how spaces become imbued with meaning. We are thus left with a static sense of social process and a somewhat gauzy view of "meaning" limited to the cosmic and the transcendent. As a result, neo-subjectivist positions are open to dismissal as overly naïve in reiterating the Cartesian cogito without an attendant sense of how spatial perception might be directed or constrained by political actions, by social inequality, by factional competition—in other words, by a wide range of social practices. The problems with neo-subjectivist approaches to space generally arise out of a rather underdeveloped account of the process of spatial production. The direct connection between culturally organized systems of belief or meaning and built space fails to take into account the social organization of production, economy, and power that allow things to get built. The house of Bachelard's daydreams (described above) is, after all, a very specific kind of house: freestanding with a cellar, an attic, and a slanted roof. This is not the row house or tenement or even the mansion or palace. It is thus situated socially in a set of political and economic relations that are largely obscured by the poetic resonance of bourgeois domesticity.

Perhaps the most troubling problem with the subjectivist position lies in its rendering of space as a purely reflective category. In dissecting the failures of spatial absolutism, phenomenological approaches make the contrary mistake of forgetting the materiality of space, its ability not only to mean but to also constrain, direct, and order physical relationships. In attending to the sensuality of space, its experience is lost. Buildings and monuments are theorized as presentations of mind, of meaning, or of memory, but they have no impact in shaping their content. As a result, neo-subjectivist approaches, particularly those in the phenomenological traditions, tend to essentialize the meaning of space. Such claims about meaning ultimately rest on what Theodor Adorno (1973) dismissively called a "jargon of authenticity," where a closed self-referential system acts as an interlocutor between form and aesthetics on the one hand and belief and cosmology on the other. In the terms of more traditional historicism, the "thou" that gained reality through a reliving of experience becomes entirely supplanted by the "I," a standpoint of unbridled egocentrism. In negotiating the distance from an apparatus of spatial perception to the expression of this spatiality in built form and back again, it is largely unclear on what basis an epistemology might rest. What bridges this distance? Lefebvre took Foucault to task on just this issue: "Foucault never explains what space it is that he is referring to, nor how it bridges the gap between the theoretical . . . realm and the practical one, between mental and social, between the space of the philosophers and the space of people who deal with material things"(1991: 4). What is missing is an account of the spatial imagination—a rendering of how "meaning" is constituted in space in the first place. As a result of the rather incomplete account of the relation between mind and form, subjectivist approaches have been overwhelmingly focused on dimensions of spatial expression where authorship is less problematic, such as architecture and site planning, to the exclusion of broader scales of spatial form and representation.

It should be noted that the intellectual lines that separate the subjectivist and evolutionist positions at times can be very thin. For example, in 1911, John Myres forwarded a distinctly historicist account of the triumph of Indo-European pastoral tribes who, forced by drought to leave their homeland, conquered surrounding peasant communities and imposed their language, beliefs, and customs on those they ruled (recounted in Trigger 1989: 168). It is not a great distance from this specific account to the general theory of State evolution offered by Carneiro where resource stress leads to warfare and the absorption of conquered communities into emergent stratified societies. The broad similarities in such accounts is not surprising, because both evolutionism and historicism are thoroughly modern ways of thinking about space. Built into that modernist foundation is the description of space as potentially indexical of sociohistorical process but certainly not of explanatory or interpretive relevance to sociopolitical order and transformation. At root, what binds together both of these modernist ontologies of space is a foundationalist impulse, a desire to hold space as a stable category of social life. If space is intelligible neither as the product of an absolute geometry nor as the imprinting of an essential human apparatus of perception, what can it be? How can it be incorporated within an approach to political life, in general, and early complex polities, in particular? The remainder of this chapter argues for a relational ontology of space, one focused on the social production of landscapes that can incorporate the experience, perception, and imagination of landscapes into a holistic, but by no means closed, framework of analysis.

Landscape and Relational Space

If space is neither an absolute category prior to all objects and thus discernible as an abstract governing geometry, nor a faculty embedded directly in the sense apparatus of subjects and thus knowable as an expression of mind or belief, then what or, perhaps better, where is it? A rejoinder to modernist ontologies of space has come in the form of a revival of a relational account, a description of space traditionally traced to Gottfried Leibniz, who framed his reply to Newton in the following terms: "I hold space to be something merely relative, as time is; that is, I hold it to be an order of coexistences, as time is an order of successions. For space denotes, in terms of possibility, an order of things which exist at the same time, considered as existing together, without inquiring into their manner of existing" (quoted in Alexander 1956: 25–26). Space, within a relational ontology, arises only as it mediates relationships among objects. Space can thus only be understood to exist insofar as "objects exist and relate to each other" (Harvey 1973: 13). The central question for analysis thus is rephrased from an attempt to assay the essential nature of space to an exploration of the practices that give rise to, solidify, and overturn particular configurations. The relational position argues that meaningful discussions of space center on relationships between subjects and objects rather than essential properties of either. Just as we discuss temporally situated events in terms of relationships (for example, the battle of Waterloo followed Napoleon's escape from Elba), so too space is only intelligible as sets of relationships.[24]

24. Two oft-cited problems with the relational account bear note but not extensive discussion. The first is the problem of unoccupied space, something Leibniz was very skeptical about. But we know that, both in common and scientific terms, it is quite possible to speak of locations that are unoccupied. As Lawrence Sklar (1974: 170–71) points out, this is not a particularly grave problem because we can describe "nonactual spatial relations." In speaking of an empty space, we are actually referring to a host of potential relationships

By refusing to hold space constant, the relational account refuses the epistemological foundations that the absolutists sought in geometric law and the subjectivists in transcendental idealism. In the absence of these anchors, we are left with an understanding of space embedded only in the practical realm of the social, a position succinctly summarized by Lefebvre's maxim "(Social) space is a (social) product" (1991: 26). To come to terms with space is to account for the relations between subjects and objects in terms of social practices. This is not to argue that the globe, the continent, the mountain, or the ocean do not exist, merely that they are unintelligible except through the social, through the ties that link subjects to objects. It is this relationship between space and practice that Michel de Certeau (1984: 102) captures in his description of space as a dimension of life that is constantly being assembled. The flow of individuals, "footsteps in the city," is what creates and defines space: "Their swarming mass is an innumerable collection of singularities. Their intertwined paths give their shape to spaces. They weave places together" and in so doing actually create the city through their daily movements and through the meanings they assign to these pathways. In other words, spaces are assembled through the collection of individuals engaged in daily practices. De Certeau describes this system of spatial production as a "pedestrian rhetoric" that composes "a story jerry-built out of elements taken from common sayings, an allusive and fragmentary story whose gaps mesh with the social practices it symbolizes."

But de Certeau's account of pedestrian rhetoric is clearly incomplete, because it is quite clear that not all individuals have the same capacity to engage in the production of spaces on the level of experience or of perception. There are constraints on the construction of landscapes, both the physical spaces and the meanings associated with them. If not everyone can produce landscapes, there is by definition a disparity in power. What makes the power to produce landscapes socially significant is that landscapes reflexively place limits on practices. Thus an ability to produce landscapes confers significant ability to influence, regulate, delimit, and control daily life.

---

that are at present unrealized. A second objection to relationalism suggests that it is incapable of providing an account of quantitative relations in space. This is simply incorrect (see Sklar 1974: 169). Most of the methods of spatial description that form the basis for cartography are based on relationships. Surveying, for example, defines quantitative relationships between points in space in terms of angles and distances. At no time does the surveyor tap into some fundamental space behind the relationships between points in space.

The proposition that not everyone is equal in their ability to construct physical spaces is self-evident. Construction of a single structure requires considerable economic resources, a significant amount of knowledge about architectural technology, recourse to a source of labor (this may or may not be an economic transaction), and permission from regulating bodies (such as a zoning board). Different projects require more or less of these elements, and thus it is not unusual to hear of individuals building a house or a barn, but it would be peculiar for an individual to build, for example, a network of roads. Hence most of the spaces that we move through in our daily lives have been built by organizations such as universities, corporations, and, most significantly and pervasively, political regimes.

The proposition that there is inequality in the production of meanings attached to particular spaces is perhaps less self-evident but is clear nonetheless. Inequality in the production of meaning has been a common theme in cultural anthropology for some time, assuming a number of forms. Sahlins has argued that "there is a dominant site of cultural production," which can be theorized as "a privileged institutional locus of the symbolic process" (1976: 211). The Marxian tradition in anthropology has more specifically located this privileged institutional locus in the apparatus of the State, describing asymmetries in the production of meaning as ideology, hegemony, and domination. In this vein, Maurice Bloch (1987) has suggested that many rituals are a mustering of meanings to support political hierarchy, and Wolf has argued that various modes for organizing labor, be they kin-based, tributary, or capitalist, "impart a characteristic directionality, a vectorial force to the formation and propagation of ideas" (1984: 398).

The problem of agency in the production of landscapes has plagued the social sciences for years; this is particularly true in archaeology, which has had an unusually difficult time envisioning actors behind artifacts. (For discussions of agency in archaeological theory, see Dietler 1998; Dobres and Robb 2000; Hill and Gunn 1977; Hodder 1986; Saitta 1994; Smith 2001.) Anthony Giddens has attempted to reconcile the opposition of empiricist and "hermeneutic" philosophical traditions and in doing so has made substantial explicit and implicit contributions to the theorization of space. He describes a reflexive relationship between agent and social structure where change in structure alters the conditions for the exercise of human action and vice versa. Paraphrasing Marx's famous dictum, he writes that "[t]he production or constitution of society is a skilled accomplishment of its members, but one that does not take place under conditions that are either wholly intended or wholly comprehended by them"(Giddens 1976: 102). What is at stake in the analysis of society is not an account of how structure, in sociological terms, determines action or how actions build structures, but rather how "action is structured in everyday contexts and how the structured features of action are, by the very performance of an action, thereby reproduced" (Thompson 1989: 56). In spatial terms, landscapes are not simply built out of a collection of practices but simultaneously constrain the possibilities for practice. By remaining within a given set of spatial parameters, practices reproduce not only the spaces themselves but also the social structures and political regimes that these spaces support. Space thus cannot be described as simply expressive or reflective, because there is much that such subjectivist descriptions leave out, exclude, or disguise. Space in this sense is recursive and instrumental, a position captured in Winston Churchill's declaration "We shape our buildings, and afterwards our buildings shape us" (quoted in Brand 1994: 3). But it is also quite true, as Stewart Brand (1994) makes clear, that this process has an unceasing temporality. Not only do the buildings shape us but we then reshape them, and the process continues both until abandonment and, as Michael Schiffer (1983) has cogently noted, long after structures become ruins and enter the province of the archaeological, rather than the architectural, record.

In moving this account of space into the social world, we emerge with the following description: space, defined as the relationships between bodies, forms, and elements, is a product of negotiations between an array of competing actors with varying practical capacities to transform these relationships. If spatial relationships are established within social practices, then inquiry must go beyond formal description to understand the physical space of the environment, the perceived space of the senses, and the representational space of the imagination as interconnected domains of human social life: "That the lived, conceived, and perceived realms should be interconnected, so that the 'subject,' the individual member of a given social group, may move from one to another without confusion—so much is a logical necessity. Whether they constitute a coherent whole is another matter" (Lefebvre 1991: 40). It is landscape that provides the conceptual apparatus for exploring the interconnections of the lived, conceived, and perceived. By embedding the production of space in the practices of actors, a relational approach demands a broad account of spatial practices. In order to understand the contests that produce landscapes, we must have some idea not simply as to how those forms organized experience but also why particular places held significance. Lefebvre suggested a three-dimensional framework for describing spatial practice based on the intersection of space with experience, perception, and imagination (1991: 38–46; see also Harvey 1989: 220–21). In the following discussion, I follow a similar tripartite framework for investigating landscapes.

DIMENSIONS OF LANDSCAPE

Experience, perception, and imagination constitute three practical dimensions of landscape. Each dimension draws on overlapping sources but demands unique epistemological positions. It is important to emphasize that these elements of landscape are separated here only for heuristic purposes and must ultimately be understood in relation to each other.

Spatial experience (material practices) describes the flow of bodies and things through physical space. This dimension of spatial practice attends most closely to distribution, transport, communication, property rules, land use, resource exploitation, and administrative, economic, or cultural divisions in physical space. Most archaeological examinations of early complex polities have been concerned with this dimension of spatial practice. Experience encompasses not only the movement through finished spaces but also the techniques and technologies of construction.

Spatial perception (evocative space or, in David Harvey's terms, "representations of space" [1989: 218]) describes the sensual interaction between actors and physical spaces. It is a space of signs, signals, cues, and codes—the analytical dimension of space where we are no longer simply drones moving through space but sensible creatures aware of spatial form and aesthetics. As Hilda Kuper noted over 25 years ago, sites, particularly political spaces, hold their potency because they crystallize evocative relationships between physical form and sense perception: "In describing 'political events,' sites such as a courtroom, a Red Square, Whitehall, the White House can be interpreted as giving an emotional effect, comparable to the power of rhetoric, to the voice of authority" (1972: 421). Although evocative space is comparable to rhetoric in its instrumentality, it is not reducible to communication with a formally constituted system of encoding and decoding. Evocative space is the analytical domain where affective terms describe interactions between humans and their environment: the dangerous spaces of alleys and docks; the inviting spaces of parks and gardens; the sterile, impassive spaces of corporate office buildings; the distinctly unmiraculous spaces of the miracle mile and the overdressed facades of the strip mall. It is this dimension of space that Susan Kus (1992) has called on archaeologists to appreciate in order to arrive at "an archaeology of body and soul."

Where spatial perception remains closely linked to form, the spatial imagination emerges entirely in discourses about space. The spatial imagination emerges most forcefully in the analytic domain of representations, from maps and pictorial landscapes to spatial theory and philosophy. This book, for example, is a representation of space, a proposal for how we should understand the spaces in which we live and the spaces that early complex polities produced for themselves. So too are the stone reliefs recovered from Assyrian palaces that portray places and events. The representation of space in these stone reliefs is no less a part of an understanding of Assyrian spatiality than the walls and doorways of the palaces in which they were displayed.

As W. J. T. Mitchell (1994: 8) has observed, discussions of space typically draw a hard boundary between the built environment and the representation of space even as the latter term evokes a teasing elision of place and image. Methodologically, this separation makes a certain degree of sense. The tools I use in mapping second millennium B.C. fortresses in the southern Caucasus, such as GPS receivers, laser theodolites, compasses, and computerized cartography programs, are of little utility in making sense of descriptions of the region found in inscribed royal annals. Unfortunately, this methodological separation has, over the past century, been reified not only into disciplinary boundaries (such as between archaeology, epigraphy, and art history) but into a body of theory that refuses to cross these lines. Yet even as textual or pictorial representations of landscape demand a unique analytical method, they are unintelligible without reference to the experiential dimensions of landscape.

The unity of experience, perception, and imagination in spatial practice demands a spatial concept that can unify the scalar divisions that have separated geographical, urban, architectural, and aesthetic dimensions of spatial analysis. The proliferation of the term landscape in the contemporary social sciences, and particularly archaeology, has prompted a growing degree of well-placed cynicism regarding the term. The concept often appears to simply denote a semantic shift—a synonym for the region, or the rural, or the vernacular, or gardens, or the physiographic environment (see Cherry, Davis, and Mantzourani 1991; Vita-Finzi 1978: 14; Yamin and Metheny 1996). If indeed landscape simply entails a concern with traditionally neglected built environments, then the term represents only a practical consideration and merits little of the theoretical gauze within which it has been wrapped. However, Carol Crumley and William Marquardt (1990) have pointed out that, beyond these highly inconsistent uses of the term, landscape provides a robust conceptual platform for integrating elements of spatial life.

Of particular import for the approach outlined here, the term landscape, with its origins in the painterly traditions, provides a conceptual framework for integrating dimensions of spatial practice across scales and media. Whereas much of this chapter has tried to break space down into analytical elements, landscape offers the chance to build space back up in close relation to a social account of temporality. As a holistic concept, landscape reminds us that dimensions of spatial practice are parts of a larger whole steeped in histories of production. These parts certainly need to be addressed but hold greater significance as they articulate with each other. Hence, an examination of pictorial depictions of, for example, gardens, holds more profound significance when integrated into a general view of the landscape, of the social organization of production, of the distribution of resources, and of the evocative sensibility of "wilderness."

TOWARD RELATIONAL LANDSCAPES IN ARCHAEOLOGY

An early movement toward a relational sense of space in early complex polities was developed by A. Colin Renfrew (1975), first as an idealized model of the early State based on Thiessen's geometry, and later as a descriptive spatial model of interpolity interactions. The former emerged from Renfrew's observation that nation-states were often preceded by clusters of "early state modules." These modules, generally identical in size and extent, were linked together by institutional and cultural features (what we might term "civilization" for lack of a better word) but were politically autonomous. In broadening this model to account for relations between polities, Renfrew and John Cherry developed a geometric model of "peer-polity interaction" where geopolitical links among polities was highly conditioned by the spatial relations between them (Cherry 1987; Renfrew 1986; Renfrew and Cherry 1986). Although the methodology used by Renfrew to detect spatial patterns in the archaeological record rests heavily on the techniques of locational geography, the goal to which the geometry of settlement is mustered is quite different. For Renfrew and Cherry, the settlement spaces of early complex polities arose in interactions within and among autonomous polities. In other words, spatial patterns are produced within and between acting sociopolitical bodies, not in correspondence to an evolutionary narrative.

A number of examinations of early complex societies, particularly studies of the Classic Maya, have implicitly adopted a relational account of space. In their account of Maya cities, Linda Schele and David Freidel (1990: 71–75) note that the production of a sacred geography in Maya urban planning and architecture materialized strategies of political competition developed over the course of several centuries. The reiteration of the primary features of the Maya mythological landscape in monumental architecture—for example, the cave leading to the heart of the sacred mountain in which grows the tree of the world—is not simply a presentation of a fully realized conceptual system. Rather, they suggest, it emerged over time as Maya rulers mapped patterns of political ritual and cosmography onto each other (ibid., 72).

Wendy Ashmore's (1989) investigations of Maya site planning can also be understood as built on a relational approach to space (see also Ashmore 1986). Eschewing the rather simplistic structuralism favored by Vogt (noted above), Ashmore argues that "settlement layouts can be rather more than passive maps of the cosmos. They can also serve as political and propagandistic tools" (1989: 272). She also argues that, through the deployment of regular site-planning templates, rulers symbolically negotiated larger geopolitical relationships, such as the identification of a lesser power with a greater one (for example, Quirigua with Tikal). Thus, a host of spatial forms are understood as negotiated in an explicitly political process. Ashmore's description of the Classic Maya built world as an instrument in the constitution of political authority is in stark contrast to Katherine Bard's recent description of mortuary architecture in Old and Middle Kingdom Egypt: "This is what the state does: it erects large monuments as symbols of authority" (1992: 5). Whereas built environments for Bard are symbols of authority, Ashmore suggests that landscapes are, in fact, constitutive of authority.

The practical implications of a relational account of landscape are extensive, particularly as they bear on investigations of political life. The relational account of space effectively removes the impassive veneer placed on space by absolutist and subjectivist accounts. Space is no longer intelligible simply as a pale reflection of transformations in political organization or expressions of politically relevant structuring concepts of the cosmos. This point is central to Lefebvre's more programmatic attempts to promote a social theory of space:

> Space is not a scientific object removed from ideology and politics; it has always been political and strategic. If space has an air of neutrality and indifference with regard to its contents and thus seems to be "purely" formal, the epitome of rational abstraction, it is precisely because it has been occupied and used, and has already been the focus of past processes whose traces are not always evident on the landscape. Space has been shaped and molded from historical and natural elements, but this has been a political process. (Lefebvre 1976: 31)

Landscapes are not simply expressions of political organization; they are political order. The chapters that follow endeavor to support this rather strong theoretical claim by investigating landscapes as constitutive elements of political life.

The relational account of space centered in the concept of landscape is one particularly well suited to archaeological investigations. Spatial relationships are the sinews of archaeological research, linking artifacts to each other, to sites, to regions, and to temporal frameworks. Although many would suggest that archaeology is the study of the past, it is in fact the study of spatial relationships between elements of material culture, some of which we extrapolate as temporal, others of which we interpret as social, political, economic, or cultural. To the extent that space is marginalized in the social sciences, so too is archaeology with its ambitious methodologies that render spatial relationships in what might seem to be rather ridiculous detail. But if space is constitutive of the social, then there might be good reason to map each artifact in its place, and archaeology, as the study of spatial relationships, might be better understood as a vital contributor to the interpretation of social life.

What is particularly interesting about archaeological investigations of early complex polities founded on a relational spatial ontology is their unconscious aversion to talking about the political in terms of "the State." The philosophical shift from modernist to relational space holds profound consequences for our understanding of political life. If space is not prior to political relationships but rather created within them, then not only must we examine spaces as political activities but we must also describe authority in terms of the spaces it assembles. Once space is understood as a set of relationships, conceptions of the political that rest on absolute foundations, such as the State, seem clunky and ill-fitting. A relational account of politics is the focus of the next chapter.

What emerges from a relational account of landscape is not an impetus to replace the temporocentrism of social evolution and historicist approaches with a new spatial centrism. Indeed, landscape does not reside comfortably as the central conceptual element of a relational sensibility. Instead, landscape emerges as a critical dimension of social practices that are themselves the proper foci of analysis. Thus the central concern of this study is not landscape but, rather, how political practices work through landscapes. A sense of the instrumentality of political landscapes requires a revision of how we conceive of the space of early complex polities as well as how we think about the State and about authority.

CHAPTER 2

Archaeologies of Political Authority

The moment we utter the words "the state" a score of intellectual ghosts rise to obscure our vision.

John Dewey, *The Public and Its Problems*

It is not coincidental that the vision of modern political analysis came to be obscured just as the spatial dimensions of human life were systematically dismissed as elements of explanation, interpretation, and critique. Only at rare moments in twentieth-century thought—when the highly abstract theoretical position afforded by the State (capitalized to reflect its universalist ambition) has receded in the face of direct accounts of the production of relationships of authority—has political life been described in explicitly spatial terms. In perhaps the most important of these accounts, Aleksandr Solzhenitsyn employed a geographic trope to describe "that amazing country of *Gulag* which, though scattered geographically, like an archipelago, was, in a psychological sense, fused into a continent" (1985: x). The significance of this rich spatial metaphor for the penal apparatus of the Stalinist regime echoes beyond its literary effectiveness. Solzhenitsyn's account suggests that political regimes must be understood in terms of the particular places that they carve out for themselves—that politics not only occupies land but also operates by and through landscapes. Furthermore, *The Gulag Archipelago* suggests that these politically produced landscapes are not conceivable solely as problems of form. The successful operation of the Gulag lay not only in the direct experience of spaces occupied by state power—a system of punishment and incarceration that stretched from the cells of the Lubyanka in central Moscow to the work camps of Kolyma in Siberia—but also in the way these discontinuous spaces were articulated in the minds of those it ensnared into a coherent landscape of authority.

The lingering importance of Solzhenitsyn's account to the way we conceptualize politics lies less in the descriptions of labor camps now abandoned to the Siberian tundra than in his exposure of the State as an illusory object of political practice, analysis, and criticism—a pretension to coherence placed on disparate and heterogeneous processes operating across a host of contiguous and noncontiguous places that together constitute political landscapes. The potency of the State as an idea arises from its representation of politics as entirely removed from space and place; the State appears complete and immune to contestation precisely because it has no geography—there is no place one can go to argue with it. Due to this lack of location built into the concept, the question "Where is the State?" though entirely reasonable, sounds quite peculiar. The only possible answer to the question is that the State is both everywhere and nowhere. The State is everywhere in that it has been implicated in every aspect of our daily lives, from the production of culture and economy on a global scale to the creation of personal identities. But the State is also nowhere. Although more than 180 political entities today are described as states, it is impossible to locate the State in the same way that we can observe governments and visit nations. From this simultaneously invisible and omnipresent conceptual location, the State provides an effective mask for political practices precisely because it obscures the inherently spatial operation of power (as an apparatus of domination) and legitimacy (as a representation of that apparatus). The State is thus a highly problematic concept for investigations of politics and one that any attempt to spatialize understandings of political life must directly confront.

Since the mid-twentieth century, the study of early complex polities has come to focus resolutely on the State. Political anthropology, the one field that has traditionally grappled with issues of political complexity beyond the narrow sociohistorical window of the modern, has centered its investigations of early complex polities on discrete political types set in universal histories. These accounts assemble various political formations into historical trajectories and thus render translucent the conditions under which polities coalesce, transform, and collapse (Adams 1966; Childe 1950; Engels 1990; Johnson and Earle 1987; Morgan 1985; Sanderson 1995; Service 1975; Steward 1972; Yoffee 1993. At the proximal end of this developmental sequence lies the State (Fried 1967; Harris 1979; Sanderson

1995; Service 1975). Although this approach has served relatively well in constructing a typology of political forms (such as Simple Chiefdom, Complex Chiefdom, and Archaic State; see Fried 1967; Johnson and Earle 1987; Service 1962; Southall 1965), outlining structural axes along which large-scale changes may occur (such as subsistence economies, exchange networks, and governmental institutions; see Carneiro 1970; Claessen 1984; Flannery 1972; Haas 1982; Wright 1977), and articulating a handful of variables (economic, demographic, ecological, and sociological) that may influence directions of long-term political transformations (see Carneiro 1970; Dumond 1972; Rathje 1971; Renfrew 1975; Wittfogel 1957; Wright 1978), it has left us with a surprisingly underdeveloped sense of what polities actually do: how rulers create subjects, how regimes ensure their reproduction, how institutions establish and defend discrete spheres of power, and how governments secure legitimacy. In short, contemporary theory has left us unprepared to describe how authority was constituted in early complex polities.

In this chapter I argue that studies of early complex polities have failed to come to grips with authority because they have over-invested theory in the concept of the State, recklessly sublimating space as a productive dimension of civil life and directing investigations toward a thoroughly illusory object of study. In projecting the State back in time and into disparate corners of the globe, archaeologists, ancient historians, and socio-cultural anthropologists have not only taken the state-illusion as reality but also bolstered the illusion by cramming a diversity of episodes of political emergence and collapse into a backstory for the modern as profound in its temporal depth as it is thin in its analytical content. If the study of early complex polities is to transform itself in ways both epistemologically coherent and critically relevant, the State must give way to studies that investigate the active constitution of political authority. Authority provides us with an account of political life rooted in concrete relationships within the civil arena (rather than in the abstract domain of universal history), and, in so doing, it grounds action, analysis, and critique in the landscapes produced within contests over power and legitimacy.

What Is the Archaic State?

In their introduction to *African Political Systems,* a canonical text in political anthropology, Meyer Fortes and E. E. Evans-Pritchard declared: "[W]e have not found that the theories of political philosophers have helped us to understand the societies we have studied and we consider them of little scientific value" (1940: 4). Yet since that time, the study of early complex polities, a research project led by sociocultural anthropologists and archaeologists, has come to focus squarely on the State, the quintessential conceptual locus of modern political theory. Although usually modified by the adjectives "early" or "archaic," the State that archaeologists assemble out of the ruins of places like the Maya lowlands or the Nile valley is conceptually the same as that employed in social and political theory. What has placed the State at the center of studies of early complex polities is a presumed historical discontinuity that divides it from previous social formations. This social evolutionary rupture is usually phrased as a "leap" (Fried 1967: 236) to statehood and bears the imprimatur of V. Gordon Childe's (1950) revolutionary model of the advent of urban societies (see chapter 5). This discontinuity is not simply envisioned as categorical—a consequence of drawing boundaries between types for purely heuristic reasons. Rather, "the Great Divide," to use Elman Service's (1975) phrase, is understood as a real historical moment. Although it is quite reasonable that some cases of political formation do represent discontinuous, radical breaks in local history (one thinks of Cortés's encounter with the Aztec Empire as one such moment), the State concept obscures the essential contingency of state formation projects. In carving off the State as a discontinuous social formation, it is reified as not simply an element of descriptive taxonomy but also a real historical phenomenon; explanation of the State is then free to drift into rival recipes in which various determinative prime movers inevitably and invariably produce a historical watershed. The State is thus intelligible only in recourse to "a style of social understanding that allows us to explain ourselves and our societies only to the extent we imagine ourselves helpless puppets of social worlds we built or of the lawlike forces that have supposedly brought these worlds into being" (Unger 1997: 7). In placing the State at the heart of investigations of early complex polities, political analysis—the investigation of the formation, administration, and transformation of civil relationships—is replaced by a political cladistics in which typological classification suffices as explanation.

THE PRISTINE AND THE ARCHAIC

In setting out a critique of the State as it has come to be deployed in studies of early complex polities, we should begin with a clear account of what is generally meant by the archaic, or early, State. Several scholars have ar-

gued that the modifiers "early" and "archaic" (treated as synonyms henceforth) restrict the field of study to those "pristine" or "primary" political
formations where the State has arisen through entirely autochthonous
processes (Khazanov 1978: 77–78). The list of primary states traditionally
includes six cases: southern Mesopotamia, Egypt, the Indus valley,[1] northern China, Mesoamerica, and the Andes. The remaining complex political formations across the world over the past 5,000 years are classified as
secondary States. The impetus to divide primary cases of State formation
from secondary ones arises from a presumption that a true understanding of the origin of the State must filter out the "noise" created by external influences. The implication is that, whereas cases of secondary formation were likely highly influenced by neighboring preexisting states
either through direct colonization or indirect effects on the economy, the
primary cases offer us an unsullied view on the most fundamental "prime
movers" of political formation at work. The pristine State is thus theorized as arising in what Morton Fried termed a "political vacuum" (1967:
231–32).

The assumption that the polities of northern China or the Andes were
somehow less influenced by their neighbors than later formations, such
as Rome or Axum, deftly combines lingering romantic views of "civilization" with modern evolutionary fantasies of the closed laboratory society. As romanticism, the idea of the pristine state insulates centers of
"civilization" from surrounding "barbarian" groups, assuming that vec-
tors of influence could only radiate from centers of incipient complexity
outward. Recent studies of culture flow between centers and peripheries
suggest that such one-way models of geopolitical influence are highly dubious (Hannerz 1992; Rosaldo 1989; Wolf 1982). The primary/secondary
divide also assumes that later States, no matter what the immediate sources
of their formation, were modeled, either positively or negatively, on preexisting ones. The romanticism of this view lies in the presumption that
a historical memory of preceding political forms or a cosmopolitanism
of neighboring organizational models is more significant to political formation than immediate conditions. Although this assumption preserves
the integrity of the State as an analytic category—through the constant
reiteration of a limited number of formal types (such as the territorial-
state, the city-state, and so forth)—the degree of mimesis in second- and
third-generation complex polities remains unassayed, and the processes

---

1. However, see Possehl 1998 for an argument against including the Indus polities in
this list.

of secondary state formation is so woefully under-theorized as to cast doubt on the utility of the primary/secondary divide.

As social evolutionist fantasy, the concept of primary states presents the possibility of a handful of historical cases where externalities are sufficiently well controlled such that conditions of study mimic the laboratory, hence the hermetic connotations of the adjective "pristine." To assume such hermetic conditions falsely demarcates early complex polities as islands, isolated from the less developed world around them. In suggesting that relations between polities constitute interference that needs to be screened out, the idea of the pristine verges on tautology because it excludes from causation a priori the effects of sociocultural exchange. It is not surprising that, in the wake of the attack on social evolutionism waged in social theory and anthropology, new attention has focused on the regional to global flow of goods, technologies, ideas, and institutions. (See, for example, the considerable debate over a proposed Bronze Age world system in southwest Asia [Algaze 1993, 2001a; Frangipane 2001; Kohl 1978, 1989; Stein 1999, 2001].)

If archaic states are not intelligible as pristine cases of complex political formations, then what distinguishes them as a discrete class of objects? Several authors have proposed that "archaic" refers to political forms based on preindustrial modes of production and so categorical distinction rests on cleaving the concept from all that came after (that is, from the modern state) rather than from all that came before (Marcus and Feinman 1998: 3; Claessen and Skalnik 1978: 4–5). However, the separation of forms of political organization solely on the basis of economic modes of production leads to significant problems in classification. Although the industrial revolution did not begin in earnest until the nineteenth century, the rise of the modern state in Europe is generally located either in the rise of absolutism or in its collapse.[2] Either way, the historical gap between the absolutist polities in Europe and the advent of industrial economies would seem to preclude founding the concept of the archaic on shifts in the mode of production. (Indeed, one might well argue that industrialism is a product of the modern state, not its precondition.)

The difficulty in clearly demarcating the archaic in conceptual terms in no way undermines the commonsense observation that certain formations arose before others and thus lend political life a temporal depth

2. Gianfranco Poggi (1990: 42) refers to the absolutist state as "the first major institutional embodiment of the modern state." In contrast, Anthony Giddens (1985: 93) argues that the absolutist polity, despite its unique features "is still . . . a traditional [archaic] state."

that can and should be investigated as historically situated processes. In this sense, the term "early" is preferred to the more evaluative term "archaic" because the former may be reduced to simply a broad temporal designation demarcating complex polities that "arose early in the history of their particular world region" (Marcus and Feinman 1998: 3). Using this sense of the term, H. J. M. Claessen (1978: 533) finds early states in various parts of the globe from the late fourth millennium B.C. (Sumer and Egypt) to the nineteenth century A.D. (Jimma [Ethiopia] and Kachari [Northeast India]). This is the sense of the term "early" to which I refer in the phrase "early complex polities" (discussed at greater length below).

THE CONCEPTUAL FORMATION OF THE STATE

Having arrived at some clarity with regard to the modifier "archaic," we are left with the much more difficult task of providing an account of the State. That we refer to the Greek *polis,* the Inka Empire, the Medieval European *regnum,* and the modern nation as all falling within the parameters of the State suggests that the term is necessarily broad and shallow in its denotation, rather than rich and deep. The State (or *status, stato, état*) is a relatively recent focus for political description and analysis. Discussions of political life in the ancient world tended to focus directly on extant forms of association. For example, although classical Greek sources often discussed the polis in general comparative terms, this was an exceedingly particular category of objects, specific to the communities of first millennium B.C. Greece.[3] For Aristotle (1988: I.ii), the polis arose through the accretion of several villages, giving it a distinctive geography composed out of a moderate to large town and a hinterland dotted with a number of smaller villages. Perfection of this form of association depended on an ideal number of citizens distributed across a territory that allowed for self-sufficiency in production, preferably with access to the sea in order to allow for long-distance trade (ibid., VII.iv–vi). The polis thus produced, as part of its fundamental nature, a unique landscape that was integral to the realization of its purpose, the good life.

Between the Aristotelian polis and the State looms a considerable con-

---

3. Despite W. Weissleder's (1978) attempt to read it as such, Aristotle's polis is not the State in the contemporary sense of the word as used by social scientists. Indeed, as Weissleder himself remarks (ibid., 201), much of the insight Aristotle provides into political organization derives from the freedom of his thought from the state concept.

ceptual distance. Whereas the definition of the polis followed from its function, the State cannot be defined by its ends. Although both conceptual objects share an understanding of political life that extends far beyond the apparatus of governance, the polis is specific and spatial, whereas the State is general and universal; where political life in the polis emerges from *relationships* among citizens, the state has come to describe an absolute phenomenon of political *form*.

An intellectual history of the State concept can be organized around four primary theoretical transformations. The first centers on the depersonalization of the State in European political discourse. Between the thirteenth and sixteenth centuries, the state was employed to denote a distinctive quality of royalty—the *status regis*—so neatly encapsulated in Louis XIV's perhaps apocryphal remark "l'état, c'est moi."[4] This sense of the term was employed through the seventeenth century both by defenders of divine right (such as Bossuet [1990: 69]) and by its critics (such as Milton [1971: 365]).[5] The extension of the state to refer not just to the quality of majesty but to the condition of the realm is known from Italy as early as the thirteenth century and from northern Europe by the early sixteenth century. In this broader sense, the state generally refers to the well-being of the political community as a whole and to the responsibility of magistrates to maintain a happy citizenry.[6] This broader view of the state is an important development because it removes the concept from an individual to encompass a geographic extent coterminous with the territory of the polity.

The second major intellectual transformation marking the emergence of the modern conceptualization of the State was its generalization into a category of political form. Use of the term "state" to describe a general type of political order can be found in a number of advice manuals for princes that appeared in the sixteenth century. The most famous among these, Machiavelli's *The Prince,* opens with a broad categorical statement: "All the states, all the dominions that have had or now have authority over men either have been or are republics or principalities" (1998: I.1–3). The state, for Machiavelli, describes the nature of the links that connect a ruler to a polity. Indeed, the substance of *The Prince* is advice on how a ruler

4. On the implications of this remark for a theory of the State, see Mansfield 1983: 849; Rowen 1961.

5. For a broader discussion of the state as a quality of rulers, see Dowdall 1923: 102.

6. In the Italian traditions we can see this sense of "state" in Giovanni da Viterbo's "Liber de Regimine Civitatum" (1901: 230), among others. A similar use of the term can be found in Erasmus's use of the term *optimus reipublicae status* (Skinner 1997: 5).

can protect the naturally tenuous ties between himself and the principality (as noted in Foucault 1979b: 8). Machiavelli's sense of state is a general categorical one; his advice draws from observations of the histories of comparable regimes and aspires to broad relevance. The state is thus generalized, divorced from the particularities of any specific polity and broadly descriptive of formal aspects of political order. It was this sense of the term that allowed Hobbes to place the concept at the center of an avowedly "scientific" exploration of public life. In the preface to *De Cive,* Hobbes codified the state as the focus of political science, defining the nascent discipline's intellectual project as a "search into the rights of states and duties of subjects" (1998: 32). But among Hobbes's immediate successors, the state remained a secondary focus of investigation, overshadowed by "nation" (as in Voltaire's *Letters Concerning the English Nation,*) "government" (as in Locke's *Two Treatises on Government,*) and "law" (as in Montesqieu's *De l'Esprit des Lois*)—concepts that emphasize the historical and geographic specificity of inquiry.

It is in Rousseau's *Discourse on the Origins of Inequality* that the state *(l'état civil)* reappears at the center of analysis, and we find it has undergone a third conceptual transformation—reification, converted from general category into a real entity capable of instrumental activity. Rousseau's state is a real phenomenon of political life, an instrument used by the wealthy to protect inequality by eroding the liberty that is the natural right of all humans. Moreover, the State is not only ceded instrumentality but also lent a universal, evolutionary history.

This universalization of the concept is the fourth and last transformation that presents us with a fully modern conception of the State (now warranting the capital "S"). Rousseau's negative understanding of the universal history of the State as a story of the fettering of natural human freedoms percolates through subsequent writings on the subject, from Marx's description of the State as a parasite on the economic order that must wither away upon the realization of communism to Freud's argument that the restrictions placed on human expression by Civilization constitute a prime source of human neuroses (Marx 1986: 322; Freud 1961). In opposition, Hegel forwarded a positive account of the State in history as the material realization of the human spirit: "In world history we are concerned only with those peoples that have formed states. For we must understand that the State is the realization of freedom, i.e., of the absolute end-goal [of world history], and that it exists for its own sake" (1988: 41). For Hegel, the State was the end of human history. Although profoundly Whiggish in its historiography, Hegel's account of the State remains foundational to explorations of early complex polities within social evolutionary frameworks.

THE STATE AND ARCHAEOLOGY

Given the close relationship between the historical development of modern political systems and the conceptual development of the State, it is not surprising that investigations of early complex polities were rather slow to adopt the concept. It is difficult to find any substantive references to the State in accounts of early complex polities prior to Lewis Henry Morgan's evolutionary cultural anthropology. Although both Hegel and Rousseau had extended the State into historiography, as an object of metahistorical process, it was only within the increasingly systematic anthropology of the late nineteenth century that the concept was deployed in a comparative study of political origins. Morgan established the state as a particular subset of government, originating in Solon's Athens and the Roman republic, founded on territorial differentiation and, more important, on political rule centered on the protection of property and organized by wealth (Morgan 1985: 338–40). Adhering rather closely to a Marxist conception, the State, according to Morgan, is simply government by and for the propertied classes, a system that he suggests continues through to "civilized society." But the State was largely secondary to Morgan's primary concern, a comparative study of the governmental institutions that mark the evolution of civilization.

With the exception of Morgan, the concept of the State is largely absent from nineteenth-century explorations of ancient societies. "Civilization"—a portmanteau of social, cultural, and political characteristics—held greater sway well into the twentieth century. Unlike the State, civilization is at its heart an evaluative term, a typological description of a society's technological, artistic, and organization achievements. As Robert Braidwood, Robert Redfield, and the *Oxford English Dictionary* all have made clear, the term "civilization" implies at its core a particular moral order, one more enlightened and refined than its dialectical antithesis, barbarism (Braidwood 1964: 137; Redfield 1953: 54–60; *Oxford English Dictionary* Online, 2nd ed., 1989). It is the tension between civilization and the insatiable drive of barbarians to blot out its achievements, a dramatic theme rooted in Herodotus and updated by Gibbon, that lies at the heart of numerous late-nineteenth- and early-twentieth-century accounts of "the rise of Civilization": "Civilization, after having maintained itself for perhaps a thousand years in extreme southeastern Europe, was thus over-

whelmed and blotted out by the northern Greek barbarians. . . . Under
the shadow of the great civilizations of the Orient, the rude Greek nomads settled down among the wreckage of the Cretan and Mycenaean
palaces" (Breasted 1919: 206). Despite the well-intentioned efforts of a
number of writers to rid the term of these evaluative implications by recasting civilization as simply the cultural matrix surrounding the archaic
State (Butzer 1980; Flannery 1972; Redman 1978; see also Scarre and Fagan 1997: 3–4), an inexpungible Victorian sensibility lies at the very heart
of the concept and so is not easily purged (as the recent use of the term
by Samuel Huntington [1993, 1996] to describe allegedly discrete and
preternaturally opposed cultural blocs makes clear; see also the more sympathetic genealogy of the term in Van Buren and Richards 2000).

Politics, generally incidental to late-nineteenth- and early-twentieth-
century studies that placed civilization at their conceptual center, was typically explored under the more restricted subheading of "government."
Inquiries into governments of ancient polities have focused almost exclusively on elucidating the formal structures of institutions, as typified
by Sylvanus Morley's (1946: 57–59) account of ancient Maya governmental organization or Samuel Kramer's (1963: 73–74) discussion of
Sumerian government. In these accounts, government denotes a specific
apparatus for administering the public order, centered in a handful of
discrete institutions. But government describes only one set of political
relationships—those securely located within discrete institutions—and thus
incorporates only a single scale of political action. Beyond the government, but within the domain of politics, lie a host of other relationships
constituted on both more expansive (geopolitical relations, territorial
sovereignty) and more refined (grassroots links to regimes, institutional
rivalries) scales of analysis.

The State began its migration into the center of studies of early complex polities in the writings of Childe, who contended that: "by 3000 B.C.
the archaeologist's picture of Egypt, Mesopotamia, and the Indus valley
no longer focuses attention on communities of simple farmers, but on
States embracing various professions and classes" (Childe 1936: 159, cap-
italization in original). However, such uses of the universal sense of the
State are rare in Childe's writings. Largely due to his intellectual debt to
Marx, Childe does not seem to have been particularly interested in the
political apparatuses of early complex polities except as they articulated
with class orders defined by relations of production. Even the chapters of
his books ostensibly focused on government move quickly to economy
after very brief definitions of forms of leadership, such as monarchical,
oligarchical, and republican (Childe 1946: chap. 10 is particularly notable

in this respect). More often, Childe employed the term "state" in a purely classificatory sense (lowercase "s"), introducing the idea of a "civilized state" to encompass several forms of political organization that followed the Urban Revolution (ibid., 155). It was in such hyphenated, categorical forms that states entered the study of early complex societies: the Classic Maya city-state, the Pharaonic regional-state of Old Kingdom Egypt, the Chinese feudal-state of the Chou dynasty (Lattimore 1940: 391; Thompson 1954: 97–98). Through the mid-twentieth century, the State remained a marginal concept in the study of early complex polities and political anthropology in general. In 1954, E. A. Hoebel summarized the rather dim view many anthropologists held toward the State, dismissing it as conceptually unrefined and analytically unproductive: "[W]here there is political organization there is a state. If political organization is universal, so then is the state" (1954: 376).

However, with the revival of cultural evolutionism in the 1950s, the State assumed central place in the study of early complex societies. Admittedly, the first-generation neo-evolutionists, most notably Leslie White (1949) and Julian Steward (1972 [1955]), were not particularly interested in the State. They were far more concerned with the operation of social evolutionary forces within groups whose subsistence economies were based on hunting and gathering or foraging. White preferred the term "civilization," most likely because it emphasized the Victorian sources (and, some might add, spirit) of his theory in the writings of Morgan and E. B. Tylor; Steward also used "civilization" rather than "State," but the former was linked to the less inherently evaluative term "complex society." It is with the second generation of neo-evolutionists that the State emerged as a central concept in investigations of early complex polities. The evolutionary typology advanced by Service (1962) placed the State at the end of a unilineal sequence that passed from Band, to Tribe, to Chiefdom. Later variants on the sequence, such as those proposed by Fried (1967) and Marvin Harris (1977) offered alternative terms for the first three stages but preserved the State as the capstone of both unilinear and multilinear metahistories.

The second-generation neo-evolutionary works of the 1960s and 1970s should be credited with placing the issues posed by complex political formations and their development squarely on the agenda of anthropology and integrating comparative studies of early complex polities within this theoretical domain. Several important volumes in political anthropology published from the 1960s through the mid-1980s are truly noteworthy accomplishments in that they integrate archaeological and ethnographic theorizations of the State into a well-developed program of research

(Adams 1966; Claessen and Skalnik 1978; Claessen, Velde, and Smith 1985; Cohen and Service 1978). However, neo-evolutionist theory must also be blamed for compressing the rich array of ethnographic and archaeological inquiries into various dimensions of complex societies, including law, government, and legitimacy, into just two canonical questions: What is the early State, and where did it come from? Before moving to a critique of the State, we must also come to some agreement on these questions.

THE ARCHAIC STATE DEFINED

Attempts to define the archaic State within political anthropology have been no less contentious than parallel attempts to pin down the State in political sociology, Marxist studies, and political science.[7] The prevailing working definitions have tended to cluster around four primary features: (1) radical social stratification preserved through (2) centralized institutions of governmental administration that (3) restrict access to and apportionment of resources through (4) the consistent threat of legitimate force.

(1) Radical social stratification has resided at the heart of understandings of the state at least since Rousseau's *Discourse on Inequality.* Stratification generally arises out of differential access within a population to limited resources that "sustain life," such as food, shelter, and the blessings of the gods (Fried 1967: 186). Differential access produces an uneven distribution of wealth. Although there is no clear standard for when concentration of wealth becomes so intense as to warrant the description "radical," archaeologists have long interpreted features such as large palaces or wealthy burials—features that highlight conspicuous consumption of goods—as sufficient markers of a social privilege far exceeding more limited fluctuations in resource distribution.

Although stratification is a generally agreed upon feature of sociopolitical complexity, there is disagreement over its causal role in the origins of the State. Both Fried and Igor Diakonoff accord stratification a central role as a "prime mover" in political evolution. Fried (1967: 230) argues that centralized government emerges inevitably out of social inequality in order to "defend the central order of stratification." Diakonoff (1991: 35–40), in a more subtle formulation, posits that unequal divisions of surpluses led to differentiation in relations of production and ultimately to

7. One study recorded 145 distinct definitions of "the State" (Titus 1931: 45), a number that has only increased in the years since.

the exploitation of an enslaved population backed by institutionalized co-ercion. Service (1975: 297–303) disagrees with such "intra-societal conflict" theories, arguing in a more contractarian vein that stratification is a col-lateral result of the functional benefits conferred by the State form. Ser-vice envisions a cost-benefit origin to the State where acquiescence to the inequalities at the heart of the early State must have been exchanged for concrete advantages, including defense from raiding parties (the State as primordial protection racket) and fostering of long-distance trade.

(2) Centralized institutions of governmental administration preserve the uneven distribution of wealth, lending stability to the State. The effec-tive autonomy of the government from the privileged social class is a mat-ter of great debate within the Marxist tradition. Some scholars have ar-gued that the governmental apparatus has no autonomy but is simply an instrument of the dominant class.[8] This direct translation of social class into political regime can be construed either narrowly, in terms of direct class control over the political agenda, or more broadly, in terms of class interests that promote and maintain advantageous forms of governance. Others have argued that governmental institutions do have some auton-omy from class structures, but both are understood as two forms taken by the same process of domination. Friedrich Engels formulated an early expression of this position:

> But now a society had come into being that by virtue of all its economic con-ditions of existence had to split up into freemen and slaves, into exploiting rich and exploited poor. . . . Such a society could only exist either in a state of continuous open struggle of these classes against one another or under the rule of a third power which, while ostensibly standing above the conflicting classes, suppressed their open conflict and permitted a class struggle at most in the economic field, in a so-called legal form. The gentile constitution had outlived itself. . . . Its place was taken by the state. (1990: 268)

More recently, Claus Offe and Volker Ronge (1997: 60) have echoed En-gels's description of the provisional autonomy of the State, describing it as an institutional form of political power that protects the common in-terests of various classes invested in reproduction of the existing order.

The central component of the State in these accounts is the way it man-ufactures sovereignty out of narrow sectional allegiances and interests. Sovereignty refers to "the idea that there is a final and absolute author-

---

8. See Bob Jessop's (1990: 25–29, 88–91) outline of the varying conceptions of the State in Marx's writings and the neo-Marxist tradition. (See also Lukes 1974; Clegg 1989.)

ity in the political community" and is generally taken as a unique feature of the State, arising in the reconciliation of a ruling apparatus and the remainder of the civil community (Hinsley 1986: 1, 22–26). Sovereignty was the implicit central feature of L. T. Hobhouse's (1911: 126–39) conception of the State as he highlighted the subordination of districts to the center and thus the creation of an institutionalized chain of command linking ruler to citizen (see also Hobhouse, Wheeler, and Ginsberg 1915). Sovereignty may be considered in personal terms, as vested in a king, or more generally located throughout the institutions of the body politic (even to its ultimate dispersal in that most democratic of institutions, the "citizenry"), but it is inherently limited geographically, circumscribed by the borders of the polity (though the inflated rhetoric of regimes may dispute the boundedness of sovereign authority).

The establishment of sovereignty through class differentiation lies at the heart of Robert Carneiro's (1970) account of the origin of the State. In Carneiro's conflict theory, the State arises out of population growth that stretches the capacities of agricultural production within a circumscribed territory to their limits. Warfare intensifies under these conditions as competition for land leads to the subordination of defeated communities. This subordination entails a loss of local autonomy, establishing the sovereignty of the victor, now ruling as an arm of a privileged class, over the entire circumscribed territory.

(3) Institutions restrict access to and apportionment of resources by asserting various degrees of control over production, exchange, and consumption. Such controls may extend from policies of tribute extraction or taxation to direct governmental ownership of key means of production and distribution. These economic policies reinforce both the centrality of governmental institutions (encouraging the expansion of bureaucracies described by Max Weber [1968]) and further concentration of wealth. However, they may also provide the basis for coalition building across social strata.

Henry Wright and Gregory Johnson (1975) have emphasized the essentially bureaucratic nature of the State, arguing that it is best understood as a decision-making organization in which specialized administrative institutions located in central places receive information from, and make decisions affecting, lower institutional tiers distributed in a network of regional and district settlements. It is not surprising that, in developing the most highly nuanced vision of the archaic State to emerge within political anthropology, Wright and Johnson's approach does not fit well within the reductionist historical vision of neo-evolutionism—the question as to why these systems of information exchange develop cannot be easily articulated with a limited set of material determinants (see, however, Wright 1978). Wright and Johnson's model calls attention to the vast networks of relationships that have been subsumed under the deceptively monolithic term, the State. These relationships link rulers to bureaucracies within institutional complexes that regulate the interconnections constitutive of the civil sphere.

(4) The consistent threat of legitimate force underlies the operation of the State as the capacity to compel adherence to the political order. Weber (1946: 77–78) was particularly emphatic in identifying "the monopoly of the legitimate use of physical force" as the defining characteristic of the State (see also the definition of the State in MacIver 1926: 22). But at issue here is not so much the actual deployment of force as the credibility of the threat. As Bob Jessop (1990: 342) has noted, though physical coercion is generally theorized as the State's ultimate instrument for compelling obedience, it has numerous other methods that it can employ, and violence is rarely the tool of first resort. What is critical to the State is not the possibility that political institutions monopolize all expressions of violence, but rather that such outbursts must be considered legitimate only in the hands of the institutionalized governmental authorities. Weber classifies the possible sources of such legitimacy as arising from tradition (belief in the historical permanency of a regime), from affect (an emotional commitment to the political order, often rooted in religious sources), from rational belief in the order's absolute value, or from belief in the legality of the order's establishment (Weber 1968: 11–12).

Stephen Sanderson (1995: 56) places a monopoly over the "means of violence" at the center of his definition of the State, suggesting that the successful suppression of a rebellion denotes a fully realized monopoly over the means of force, a power that distinguishes the State from previous political formations along the evolutionary progression, such as the Chiefdom. In addition to the arguments by Carneiro (1970) and Giddens (1985) that most States, ancient and modern, do not hold such a monopoly, Sanderson's definition has the added drawback of precipitously reducing the State to a synonym of repression.

Numerous other features have been prominent in definitions of the State, including administrative technologies (such as writing) and "ethnic stratification." (On the former, see excellent articles in Marcus 1992; Michalowski 1987; Postgate, Wang, and Wilkinson 1995; Rothman 1994; Schmandt-Besserat 1996. On the latter, see Thurnwald and Thurnwald 1935; rejected in Cohen 1978: 32.) Similarly, a number of other "prime

movers" to State formation have been identified, including the managerial demands of irrigation and interpolity exchange (Polanyi, Arensberg,
and Pearson 1957; Renfrew 1975; Wittfogel 1957). However, the definition
of the State discussed above pulls together enough of the main elements
to serve as a fair description of the concept and as a locus for critique.

It should be made clear that the State concept, as presently constituted,
aspires to describe both a real phenomenon of political life and a general
type of political organization. This is possible because of the essential unity
attributed to the concept. The assumption of a phenomenal singularity
underlying the diversity of political forms was defended in 1927 by
Robert H. Lowie, who proposed that the State is a form of association
whose expression in diverse global contexts is a result of the psychic unity
of mankind.[9] Although the psychological locus of Lowie's unity postu-
late was dismissed during the materialist turn in archaeology, it survives
today rooted in a simplistic biology where it has been asserted that the
lack of speciation among human groups provides the basis for a unified
story of cultural transformation (Mann 1986: 34–35). In preserving a pos-
tulated unity, a particular epistemological stance was also preserved that
assumed differing features of complex societies to be epiphenomenal,
whereas shared features reflected the "kernel" of the State. This is, of
course, a tautological enterprise; the shared classificatory features that are
labeled "the State" are also held up as proof of the State's reality as a generalizable social fact. However, it allows the State to have not only the
status of heuristic category but also the ambition to describe a phenomenon of political life that offers a more encompassing and compelling locus
for political analysis than, for example, government and law—concepts
that are marginalized as particular and epiphenomenal rather than universal and explanatory. However, even as the State is transformed into a
real element of various public and private relationships, it becomes increasingly difficult to locate because, though the State has been given temporality, it has no place.

Against the State: An Archaeological Critique

Having developed a picture of the State and its deployment in theories
of the evolution of early complex polities, I want now to discuss four pri-

9. This psychological basis for presuming essential unity was later folded into a social
evolutionary approach in Krader 1968: 11–12.

mary reasons for abandoning the concept as the centerpiece for political analyses of early complex polities. The first reason centers on the denotational insecurity of the term.

DENOTATION

The State, despite its centrality, is an entirely nebulous object of study without a clear referent. Fried began his landmark *The Evolution of Political Society* (1967) by bemoaning the horrors of definition and highlighting the pointlessness of it all if we cannot agree on common denotations as tools of analysis. But Fried's complaint was already quite dated in respect to the State. Bolingbroke decried the instability of any definition of the State in the eighteenth century, writing that "the state is become, under ancient and known forms, an undefinable monster" (quoted in Skinner 1997: 19), and in the mid-twentieth century David Easton (1953: 107–8) led a functionalist assault on the concept's lack of precision. Even Weber's oft-cited definition of the State as "a human community which (successfully) claims the monopoly of the legitimate use of physical force within a given territory" (1968: 78) quickly disappears from view, obscured by numerous qualifications on the nature and deployment of force, the variable forms of legitimacy, and the meaning of community that shift analysis away from the State and toward bureaucracy (see also Jessop 1990: 343). In the most thorough recent examination of the State from a Marxist point of view, Jessop demurs on offering a definition until his final chapter and follows it with a long list of caveats. Pierre Bourdieu recently formalized this tradition of denotational insecurity, arguing that the state must be subject to "a sort of *hyperbolic doubt*. . . . For, when it comes to the state, one never doubts enough" (1999: 54).

It might be objected that, though the multivocality of landscape was applauded in the previous chapter, I am here condemning the denotational plurality of the State. In response I suggest that the difference between the two cases lies in the analytical work that can be accomplished by the respective concepts. Whereas investigations of landscape have flourished as the meaning of the term has expanded, such conceptual fuzziness is extraordinarily problematic for accounts of the modern state. So much effort has been spent over the past 30 years in trying to define the archaic State that classificatory debates have sapped much of the field's theoretical energy. The result has been an extended period when categorization and classification have substituted for analysis. For years the primary tension in this endeavor was provided by a strangely vituperative dispute over the categorical and evolutionary division between the Chiefdom and the State (see, for example, Bawden 1989; Carneiro 1981; Earle 1987, 1991; Kirch 1984; McGuire 1983; Wright 1994; Yoffee 1979, 1993). Generally lost in this struggle was a concern for the actual dynamics of polities.

Despite the often grand claims and bitter language, the intellectual stakes in these classificatory debates are extraordinarily low. That is, no great insight into the nature or origins of early polities could issue from their resolution. Classification of polities as State or Chiefdom is singularly uninteresting because it speaks only to failures in our typological imagination, telling us little about the nature of politics in the past. The same shift in focus from what happens in ancient polities to the variability among and within types now threatens to engulf the analytical concept of the city-state. In a recent volume on the city-state as a "cross-cultural regularity," Norman Yoffee argues that "we must resist elevating the [city-state] to an intellectual fetish. We must unpack the term *city-state*, trace the variability within it, and explain the significant divergences from it" (1997: 256). Despite the well-taken caution against fetishization, rather than reducing the city-state to a simple heuristic category, a creation of our own typological imagination, Yoffee goes on to position the concept as a real product of social evolution, a social form that lies at the terminus of various trajectories of mounting social inequality: "In general, city-states evolve as collecting basins for the crystallization of long-term trends toward stratification and social differentiation in a geographical region" (ibid., 261). In taking the city-state as a real, singular, historically stable form of political association, this new social evolutionary program threatens to merely reinstantiate the existing theoretical emphasis on categorization in only slightly different terms. The problem lies not in typology per se but rather in mistaking these analytical types as enduring spatial forms with their own historical inevitability.

EPISTEMOLOGY AND ILLUSION

A second argument for abandoning the State arises from epistemological sources. A growing number of critics have attacked the State's pretension to describe a coherent object of analysis, a real agent, or a site of political action. Timothy P. Mitchell has argued that the State cannot be understood as an object, "a free-standing entity, whether an agent, instrument, organization or structure" (1991: 95). Rather, the state (lowercase "s") can be understood to emerge as "a set of powerful yet elusive methods of ordering and representing social practice" (Mitchell in Bendix et al. 1992:

1017). The goal of Mitchell's Foucauldian critique is to break down the conceptual boundaries of the State that divide it, as an object of inquiry, from the larger social world. Without its conceptual discreteness, the State merges into the broader field of social practices such that investigation of the political is transformed into an investigation of "structural effects" rather than structures, "of detailed processes of spatial organization, temporal arrangement, functional specification, and supervision and surveillance" (Mitchell 1991: 95). The state endures, in Mitchell's analysis, only as a historically specific political formation, one unique in its current incarnation, where the illusion of a separate political entity dictating policy for society holds profound consequences for action.

Philip Abrams has forwarded a similar critique in a more sociological vein that owes less to Foucault than to Marx. The State, Abrams argues, is not a real dimension of political life but, rather, represents a fetishization of twentieth-century political ideology as deep metahistorical structure: "The state is at most a message of domination—an ideological artefact attributing unity, morality and independence to the disunited amoral and dependent workings of the practice of government" (1988: 81). The State is thus reconsidered as a fiction created through the mistaken reification of a classificatory type. The term unifies and gives conceptual coherence to what are in fact a large number of discrete political practices. The State, Abrams continues, is not the unified locus of political life that we have assumed but simply one of the many masks worn by authority in order to pretend to be universal, inevitable, and sewn in to the very deepest fabric of the way things are.[10] As he concludes: "[T]he state is not the reality which stands behind the mask of political practice. It is itself the mask which prevents our seeing political practice as it is" (ibid., 81). What better guise could inequality, domination, and exploitation take than to be represented as part of a metahistorical order that is simultaneously everywhere and nowhere?

The epistemological consequences of Mitchell's and Abrams's critiques are quite staggering for the study of the archaic State. To the extent that they have attempted to detail the origins of the State, studies of early complex polities have been engaged in writing a profound backstory to current politics, one that universalizes and thus legitimizes current political systems by rooting them in a far-off antiquity connected to the present in an unbreakable chain of historical causation. By reading the State out

10. For an extended discussion of critiques of the State as illusion, see Comaroff 1998.

over five millennia, archaeologies of early complex polities contribute to a crisis of faith in the viability of contemporary political action; by rooting sociopolitical organization in metahistory rather than the immediacy of landscape, social evolution suggests that the alteration of present conditions can come at best only through full-scale revolution (the vision of Marx) or at worst cannot (and should not) come at all from within the social world (Unger 1997: 6). Such inevitability can only lead to a debilitating resignation from the political arena.

CRITICAL PRAXIS

This leads us to the third problem with the concept of the State: its questionable capacity to support critical reflection generative of contemporary praxis. It is unclear what immediate practical significance the State holds in the post–Cold War era. In 1919, Lenin was able to credibly assert that the State had become "the focus of all political questions and of all political disputes of the present day" (Lenin 1965: 19). That political disputes should center on differing conceptualizations of the State was central to revolutionary Bolshevism in that the traditional Aristotelian arguments over the proper form of government—republican, monarchical, aristocratic—were transformed into a much broader debate over the history and future of sociopolitical formations.[11]

Considerable import rested on defining the nature and origins of the State during the Cold War, and extensive resources of both the Soviet Union and the West were put into anthropological and archaeological research that aspired to explicate the true nature of the State in both specific (structural) and general (evolutionary) terms. At the height of the Cold War, what we might call the conservative evolutionism of Service contrasted with the liberal and Marxist models forwarded by Fried in the United States and Diakonoff in the Soviet Union.[12] One can see the conflict between these accounts as, in a highly abstracted sense, a search for the historical legitimacy of competing conceptions of the State. Service's call to forward studies of the State by setting aside "as irrelevant our po-

11. As Hayden White (1973) has pointed out, we can see in the heirs of Marx and Tocqueville not simply rival concepts of proper governance but also rival emplotments of the past and future of society—a contest played out in part in the intellectual terrain defined by the evolution of the State as inscribed within studies of early complex polities.

12. See Lenski 1966 for a discussion of the differences between radical and conservative approaches to inequality and Kehoe 1998 for an excoriating history of the politics of the New Archaeology.

litical views about our present state" (1975: 289) not only comes across as a tad disingenuous (the preceding paragraphs had ridiculed the Marxism of C. Wright Mills, leaving the equally problematic thought of Kingsley Davis unchallenged) but also somewhat undesirable; it was the amplification of these debates in reference to ancient societies that made them relevant to contemporary political discourse. In particular, the debate among Marxist, liberal, and conservative writers over the role of class stratification in the emergence of the archaic State, though always cloaked in the folds of scientific argumentation, was a vital extension of the study of early complex polities into contemporary political discussions.

It is difficult to see how such a debate might inform critical thought regarding the current situation of the world. If Lenin's remarks in 1919 mark the emergence of the State as the pivotal concept of critical political historiography, then the concept's fall might appropriately be located in comments made by Mikhail Gorbachev at the United Nations 70 years later: "We [the Soviet Union] are, of course, far from claiming to be in possession of the ultimate truth" he declared, implicitly deflating Marxist claims to superior knowledge of human political history and mechanics (quoted in Nelson 2000: 4). In today's post–Cold War world, very little hangs in the balance in competing conceptualizations of the State. Some might fairly argue that contemporary arguments between defenders of the welfare state and advocates of neo-liberalism are substantially invested in opposed understandings of the State. However, I would argue that these are conflicts over policies and practices within rival visions of liberal governance, not over competing claims on the essence of the State. That is, their differences are not, by and large, predicated on competing theories of history but on more or less technocratic accounts of proper governance.

Events have clearly outpaced the intellectual apparatus through which the study of early complex polities can be deployed to allow the past to provide illumination on the present. The myriad conflicts that now grip the world reveal not the practical irrelevance of the nation-state, as some authors have argued, but rather the conceptual obsolescence of the State concept itself as the orienting object of political research. Although clearly great political struggles are ongoing, can debate over the nature or origins of the State shed any illumination on conflicts in southeastern Europe or the rapidly increasing chasm between rich and poor in Western democracies? I think it unlikely. The very features that made the State compelling for Cold War rivalries—the eschatological view of history rooted in remote antiquity—make it appear pious and impractical today. What seems more clear today than ever before is that politics is less about rival metahistorical vision than it is about rivalries over historically situated landscapes.

ONTOLOGY AND SPACE

A final difficulty in building political analysis on the State is ontological and centers on the aspatial nature of the concept. Though argued from the point of view of landscape in chapter 1, the exclusion of space from State theory is also objectionable from a political perspective. A spatial critique of the State has been forwarded in a number of disparate arenas. Immanuel Wallerstein's World Systems theory reinvigorates a geographic understanding of politics by according central analytical position to relationships established in space between economically intertwined places (Wallerstein 1974; see also Johnston 1982: chap. 4; Taylor 1997). In Wallerstein's account of the modern World System, the rise of European (later American) industrial capitalism was predicated on the creation of a set of economic relationships between the centers of industrial production—the factories and mills of Europe that transformed raw materials into finished products—and the peripheral regions that served as sources of raw materials and markets for re-export of finished products. The particular significance of the World System lies in the instrumentality accorded to geographic position. The relationships established between center, semiperiphery, and periphery are central to formation of political regimes and political economies because it was relations across space—and the transportation technologies that facilitated them—that enabled Europe to build itself on the production of goods from raw materials dispersed around the globe.

It is important to note that, as originally conceived, the World System not only described a particular set of places but was also set in a very specific time: the development of capitalist economies since A.D. 1500. Attempts by archaeologists and historians (and indeed by Wallerstein himself) to read a World System back in time, into the era of early complex polities, tend to obscure what is most interesting about the model (Algaze 1993; Blanton and Feinman 1984; Champion 1989; Eckhardt 1995; Frank 1993; Kohl 1989; Stein 1999; Wallerstein 1995). The explanatory power of the World System model lies in its confident intersplicing of variation across both space and time. To move it out of the modern age requires the crystallization of the complexities of its geography into rigid generalities (as evidenced by the way archaeologists tend to unceremoniously dump the semiperiphery from the model with little discussion or justification). Hence the World System taken outside of the modern context generally depends on reestablishing an absolute spatial ontology founded on a limited set of highly crystallized geographic positions. The lesson to be drawn from Wallerstein's model is that considerable explanatory power and interpretive elegance can arise from accounting for social, political, and economic complexity in terms of the spatial relationships they produced. These relationships, as Gil Stein (1999: 6–7) has pointed out in his thoughtful critique of the use of World Systems models to describe fourth millennium B.C. southwest Asia, are exceedingly heterogeneous depending on both local and regional conditions.

A second prominent source of spatial critiques of the State can be traced to several studies of modern urbanism penned in the 1970s and 1980s. With roots in David Harvey's *Social Justice and the City* (1973) as well as Henri Lefebvre's *La Production de l'Espace* (1991), the fluorescence of this criticism of the State through the filter of the city can be found in Manuel Castells's seminal study, *The City and the Grassroots:* "Therefore spatial forms, at least on our planet, will be produced by human action, as are all other objects, and will express and perform the interests of the dominant class according to a given mode of production" (1983: 311). The key terms in Castells's declaration are "express and perform." Spatial dimensions of political life not only reflect aspects of structure but are also instrumental in producing power and legitimacy. This account of political space—a vision ultimately quite close to that offered by Solzhenitsyn—has been powerfully developed in a number of geographic and architectural studies, from Mike Davis's (1990) searing indictment of Los Angeles urban development as a tool of political exclusion and domination to Sharon Zukin's (1991) nuanced exploration of the interplay of spatial production, political institutions, and social fantasies in the creation of the postmodern American landscape.

Within this tradition of political criticism, the State is maddening in its lack of place, obscuring the intense everyday political concern with ordering landscapes. In practical terms, the lack of any location for the State makes political action almost impossible. In rallying against the State, where must one strike? In coming to the defense of the State, where does one rally? On the most fundamental level, the State fails to adequately apprehend political life because of its a priori demotion of space to the status of epiphenomena. The central reason for this dismissal is that the State is built on both an absolute ontology of space and an absolute ontology of politics. Civil society rests in a limited repertoire of possible forms that arise and transform in predictable (or at least retrodictive) ways because they follow a rather limited set of rules. But, like space, politics arises in relationships between groups and individuals, not full grown from a repertoire of types. This relational account of politics begs numerous questions—what is the nature of civil ties? what kinds of groups or subjects should be thought of as agents within the political sphere? what sociocultural links parallel political relationships? These issues, I suggest, provide far richer foci for investigations of early complex polities than debates over typological assignments.[13] A relational conception of political life must therefore be built not on the rigid absolutism of the State but on authority, an umbrella term for a set of concepts that emphasize the intersection of space and time in political practice.

What Was Political Authority?

I have argued above against a particular concept of the State, one that vests all political action and explanatory power in a universalized set of structurally discrete formations. But where does the dismissal of the (archaic) State, as a real object of inquiry, leave archaeological and historical analyses of political life? What conceptual apparatus can we build to re-center analysis on what polities do rather than what type they resemble? If we throw out the State as the central object of study, studies of early complex polities must confront two separate problems that have traditionally been conflated. The first problem is to identify the class of objects under examination and define the heuristic bases on which they are ceded a degree of analytical discreteness. The second problem is to identify a productive conceptual locus for investigations of the political. That is, what conceptual foundation can provide a basis for understanding the relationships that constitute the political sphere of action?

EARLY COMPLEX POLITIES

The first problem is typological and thus of limited theoretical consequence. The rejection of the archaic State leaves us with the much more grounded, and consequently less grand, term "early complex polities" as a provisional description of the objects of comparative inquiry into the

---

13. It is important to note that, even if we abandon the State as the conceptual locus for detailing politics in early complex polities, an imperative remains to account for the concept's modern ideological construction and its rhetorical hegemony. This work, unfortunately, extends well beyond the limits of the present discussion.

historical roots of contemporary modes of political action.[14] Where the archaic State was amorphous and difficult to define, the early complex polity, though not unproblematic in itself (as I discuss in the last chapter), is clear in its terms yet flexible in its connotation—an advantage because there is little likelihood that much debate will be stirred over whether particular formations can be described as early complex polities. The intellectual stakes involved in such a designation are simply far too low. It is important to emphasize that political complexity—a means of referencing certain sociological features such as inequality or centralized organs of governance—does not equate with cultural complexity, which tends to center on an evaluative positioning of social groups along broad trajectories of social development.

The modifier "early" refers to a temporal horizon relative to local histories of political transformation. What distinguishes early complex polities from more recent formations is not a profound historical divide but merely a methodological shift. The study of early complex polities is vested as profoundly in archaeology as it is in history and epigraphy. The relative primacy (or at least equality) of the material culture record sets "early polities" apart from later formations where the historical record tends to swamp archaeological studies.[15] This by no means restricts the field to nonliterate societies; the historical records of, for example, Mesopotamia, the Classic Maya, or the early Chinese dynasties are only intelligible in connection with the archaeological contexts of their discovery. Furthermore, these documents tend to be so restricted in their concerns that they cannot provide sufficient bases for understanding social life without archaeological research. As a result, early complex polities demand a unique epistemological stance that makes use of the extant archaeological and textual records even as they are historically continuous with more modern formations in which studies of material culture play an important, if somewhat different, part.

The term "complex" typically refers to a very general set of sociocultural features, such as inequality in access to resources, variability in social roles, differentiation of decision-making bodies, permanence of institutions, and

---

14. A number of other archaeologists have also eschewed "the State" in favor of variations on "early complex polity." Olivier de Montmollin (1989) adopts the synonymous term "ancient complex polities" in his study of the late/terminal Classic period in the Rosario Valley of Mexico.

15. This does not imply a diminution in the importance of archaeological knowledge to understanding more intensely textual eras. Rather, I am simply pointing to an epistemological shift that necessarily follows the incorporation of a wealth of written sources.

the distribution and flow of symbols, meanings, and practices (Blau 1977; Hannerz 1992; McGuire 1983; Tainter 1988: 23–31). In contrast to the State, a noun with pretensions to represent real structures, "complex" is an adjective describing the *relative* extent, heterogeneity, and differentiation of sociocultural formations. However, I do not wish to suggest that complexity is an entirely unproblematic category. As Randall McGuire (1996) has cogently argued, the term can obscure the operation of a wide range of social phenomena by reducing them to a rather narrow set of social operations—a problem most manifest in the strident reductionism of the so-called "complexity" theorists of the Santa Fe Institute (see also Gumerman and Gell-Mann 1994; Waldrop 1992). Moreover, complexity does not entirely satisfy as a value-neutral adjective for differentiating qualitatively different objects of anthropological analysis. For this reason, the term has undergone what Yoffee refers to as "complexity inflation" as even social communities from the Paleolithic have come to be described as "complex" (personal communication). However, this need not be a problem so long as we are clear about what dimension of complexity provides the focus for any given study.

A number of political scientists, anthropologists, and sociologists have strongly contested efforts, such as those made by Michael Mann, to carve out a structurally discrete political sphere (Bourdieu 1999; Corrigan and Sayer 1985; Mitchell 1991). Politics intrudes far too profoundly on economy, culture, and social life to be confined within a well-bounded institutional position. However, we can describe the political in terms of practical relationships that are strongly shaped by public forms of civil action. I suggest that four relationships in particular must lie at the heart of how we conceptualize politics:

- interpolity ties, or geopolitical relationships;
- relations between regimes and subjects that forge the polity;
- ties among power elites and their links to grassroots social groups (such as kin groups, occupational associations) that constitute political regimes; and
- relationships among governmental institutions.

All four of these sets of relationships can be described as strongly political in that they have immediate and profound effects on the life of the polity. These relationships by no means exhaust the political, but they are central to it. Thus, the political here describes a flexible set of public relationships that organize practices of domination, governance, and legitimation. In other words, what is at issue in an examination of politics in early complex polities is the constitution of authority.

ON AUTHORITY

Authority is a neglected concept not only in the study of ancient societies but in modern ones as well. A prominent reason for this lacuna is the political theory of Hannah Arendt (1958), who forcefully argued that authority first entered political life during the Roman era only to disappear again during the twentieth century. Arendt's argument has been reiterated in conservative intellectual circles since the 1960s as an indictment of leftist assaults on then-dominant institutions. As a result of Arendt's historical positioning of the concept, authority was defined as of little concern to political analyses of either modern or ancient polities. Arendt's argument was that authority is only possessed by a regime that faithfully transmits the principles of a society's foundation through succeeding generations. In the case of Rome, the authority of the living derived from the sacred roots of the founders transmitted from Aeneas and Romulus to successive generations of elders institutionalized in the Senate.

The most immediate objection to Arendt's account of authority is situational. The context of Arendt's writing is important to elucidating the historical peculiarity of her position. As Bruce Lincoln (1994) has outlined at some length, Arendt's argument arose out of two sources. In grouping both Nazi Germany and Stalinist Russia under the general category of totalitarian regimes, Arendt's seminal work *The Origins of Totalitarianism* (1951) was profoundly invested in connecting communism with the dangers of fascism that had mobilized the United States in World War II. This intellectual position, supported in part, we now know, by covert monies from U.S. governmental organizations including the Central Intelligence Agency, attempted to rebut an analytic position forwarded by Theodor Adorno and other members of the Frankfurt School. Adorno's research in the United States had suggested that many traits of the "authoritarian personality" that led the German population to embrace fascism were highly prevalent among Americans as well. Such a focus on authoritarianism threatened to shade the sociological and moral divide between the Soviet Union and the West at a time when various groups clustered around the anticommunist banner sought to portray the contrast in stark terms. Today, it is difficult to share Arendt's concern that authority be so conceptually reduced.

This brings us to a second objection to Arendt's account of authority. In his analysis of authority, Weber outlined three basic forms—traditional, charismatic, and rational-legal—that could be differentiated on the basis of "the kind of claim to legitimacy typically made by each" (1947: 324). Rational-legal authority rests on a belief in the legality of rules rooted in a general confidence in the procedural validity by which they were enacted. Charismatic authority arises in the devotion of a population to the "specific and exceptional sanctity" of an individual and the rightness of any social order ordained by him or her. Traditional authority is grounded in the belief that the present order rests on a sanctified politicocultural history. It is clear that Arendt wanted to exclude the first two forms and limit authority to the traditional. This move is unconvincing, however, because Arendt never considers how tradition, charisma, and rational-legal forms of authority might grow out of and into each other over time. The transition from the Roman republic to the empire under the Caesars can be seen as just such an elision of tradition and charisma.

The rigidity of Arendt's sense of the term is largely due to her description of authority as an entity to be possessed. As Lincoln has persuasively argued, authority is not an entity but, rather, is best described "in relational terms as the effect of a posited, perceived, or institutionally ascribed asymmetry between speaker and audience" (1994: 4). As a result, authority is not an obsolete element of political life but is, and has always been, a fundamental element of civil discourse. Lincoln's erudite analysis of authority in contexts ranging from the Homeric Assembly to contemporary environmental protests indicates that the relations constituting authority emerge primarily in linguistic performance. Authority, he argues, "lies in the capacity to produce consequential speech" (ibid.). What is said, Lincoln notes, has little consequence. What is central to authority is the effect produced by speech. Authoritative speech reassures the doubtful, wins over the ambivalent, and defuses the opposition.

Although claiming precedence for the operation of authority through language, Lincoln notes the importance of nonverbal dimensions of human intercourse, including landscapes, material culture, gesture, and costume. But he dismisses these as gimmicks—the smoke and mirrors that help "bamboozle" an audience but are ultimately secondary to the show carried out in language (ibid., 5). David Bell goes even further than Lincoln, arguing a neo-Aristotelian position that "politics is talk" (1975: 10). If politics were indeed only talk, it would be rather difficult to make sense of a wide range of contemporary political activities from taxation and war to the protest march and sit-in. In such fields of explicitly political practice, talk is rather beside the point if not practically impossible. Politics certainly involves speech, but it would also seem to be, in a more encompassing sense, about human action and movement, and hence about landscape.

There are good reasons to think that language is only one part of the larger constitution of authority through practice. Let me use the following example. A judge in a court of law is said to have authority in that court. Outside of that place, though possessed of considerable status (suggesting a great deal of transitivity between status and authority), a judge's speech is not authoritative because he or she cannot simply condemn passersby to jail sentences. Likewise, another individual who sits on the bench and pronounces judgments is at best recognized to be a fool and at worst a usurper. Either way, such pronouncements are not legitimate and thus not authoritative. The relationship between people and places is quite clearly an important constituent of authority.

The speech of the judge cannot, I think, be said to create authority, because that must already exist if his words are to be distinguished from those of the usurper. Indeed, the words exchanged in a courtroom are only significant if the authority of the proceedings has *already* been established, before the first word is uttered. This throws our analysis of the roots of authority back on the space of the court itself. As the example of the usurper shows, authority cannot be said to inhere in space itself but rather in the relationship between the place and the actors. The constitution of authority therefore hangs on the production of place as it organizes and conditions the actors who will practice through, rather than simply within, specific landscapes. Authority is constituted in the court through the production of its space providing an arena in which speech by designated actors can already be constituted as authoritative. Authority is thus not a favored child of language but an effect of relationships assembled and reproduced in the much broader realm of political practice, in which landscape plays a central role before the first authoritative word is uttered. As a result, we should look for authority not simply in the relatively rarefied realm of a handful of literary texts but on the ground, in the archaeological record.

A large number of human relationships may be described as authoritative: parent-child, teacher-student, boss-worker, and the like. Ethnographers have detailed the constitution of and challenges to authority in contexts as disparate as the cadres of rural China and the punks of Brixton (Emery 1986; Siu 1986). But the analytical importance of explicitly *political* authority lies in its presumptive claim to be the authority of last resort, able to exert its commands within all other such relations and thus reconfigure them, if only momentarily, in the public realm. Political authorities can come to the defense of children against parents and workers against bosses. The central question for the study of early complex polities is thus not the origin and evolution of an essentialized totality that we call the State but an inquiry into how, in varying sociocultural formations, an authoritative political apparatus came to gain varying degrees of ascendancy over all other social relations.[16] This is what we *should* mean when we refer to states, if the concept is to have any utility: those polities where a public apparatus holds the legitimate power to intercede in other asymmetric relationships in order to mark itself as the authority of last resort.

Authority can thus be located in the confluence of interests that promote fundamental practical asymmetries and embrace the paramountcy of the resultant apparatus as inherent to the proper order of things.[17] At the center of this understanding of authority rest two fundamental relational processes: the power to direct others, and the recognition of the legitimacy of these commands. Studies of political organization in anthropology have resolutely moved to consider the concept of power in ethnographic and archaeological contexts under the influence of theorists such as Mann (1986), Michel Foucault (1979a), Steven Lukes (1974), and Nicos Poulantzas (1973). Indeed, the concept has become so ubiquitous that it has, in many cases, come to exhaust analysis of processes of political formation (Haas 1982). For the purposes of this study, power is not just the ability of one (individual, class, regime, polity) to realize its (political, economic, social) interests at the expense of another (individual, class, regime, polity) but, more profoundly, the capacity to constitute interests and determine their significance within the management of existing conditions.

As important as power relations are, they are only one aspect of authority. As Weber argued, power may ultimately be exerted in similar ways in varying forms of authority; it is the bases of legitimacy that differentiates them. By legitimacy I refer to the ability of a regime to synchronize practices that perpetuate the existing political order within a discursive framework that generates the allegiance of subjects. This description coin-

16. See Shamgar-Handelman and Handelman 1986 for a study of how educational institutions in Israel try to set the nation-state (rather than familial structures) at the apex of authority.

17. See also the definition of authority offered in Ferguson and Mansbach 1996: 35.

cides generally with Antonio Gramsci's (1971: 182) description of the third moment in the historical relation of social forces in which the interests pursued by a regime are perceived by subjects to coincide with their own interests. Such a view of legitimacy also generally accords with John Baines and Norman Yoffee's (2000: 15) definition of the term as "the institutionalization of people's acceptance of, involvement in, and contribution towards order," although I would place particular emphasis on the resolution of the oppositional nature of power through a redefinition of the interests of subject groups.

Both power and legitimacy are necessary to constitute authority. A legitimate government in exile holds no power and thus cannot be said to be in authority; conversely, a regime without legitimacy, based solely on domination, may be described as piratical or extortionist, but not authoritative. In discussing the constitution of political authority, I am referring to the process by which relations of political power emerge within a framework of legitimate governance. This process of constitution is productive in that it emerges, as Giddens (1984) has outlined, in the reflexive interrelations between historically situated agents and existing structures.

Space and Reproduction:
The Temporality of Political Landscapes

Landscape envelopes both the spatial and the temporal. Although the following studies emphasize the intertwined historicity and spatiality of political production and reproduction, it is worthwhile to briefly outline here just how one might formally theorize the temporality of landscape. That is, if authority is an effect of spatialized practices, then we must pose the question as to how authority waxes, wanes, and reproduces over time. What is it that generates the need for regular reproduction of authority once it is understood as practice closely linked to the production of space? We can identify four primary sources of emergent gaps between authority and landscape that generate a demand for constant reproduction (as well as providing potential locales of strong critique). The first and most important arises from the constellatory nature of the political landscape. As Solzhenitsyn suggested in *Gulag Archipelago,* the pretensions of regimes to geographic universality—to an immanent presence—is an illusion cloaking what is at best an archipelagic landscape created in practices of governance and oversight of varying intensity and sustainability. Thus, the political landscape's inevitable incompleteness sustains a relentless effort to fill in absent positions and re-map the imagination of the polity.

A second source of the need for reproduction arises from the unique nature of the built environment as an artifact. The political landscape as a material instrument in the constitution of authority emerges as not only incomplete but also highly inadequate. Once something is built, it immediately begins to age at a rate wholly incommensurable with the endurance of various policies and strategies for rule. Thus political authority exists in constant tension with the very landscape that generates and sustains its rule.

A third drive to political reproduction arises in the incomplete relationship between land and landscape. Environmental transformations such as major droughts and shifts in river courses can have a profound impact on political orders, necessitating wholesale transformations in landscapes, both physical and representational. As Davis cogently argues in *Ecology of Fear* (1998), it is not the environmental catastrophes that constantly strike modern Los Angeles per se that provoke shifts in relationships between political regimes and constituencies. Rather, it is the imagination of these landscapes on which effective political coalitions are built.

A fourth source of the need for reproduction arises from the fractured nature of authority. Where the State presents us with a rather singular totalized view of the political, authority is more broadly distributed in a wide array of relationships. Landscapes are produced in many different spheres of authority (for example, families build houses, corporate groups construct settlements, rulers establish cities). As a result, the landscape emerges as a palimpsest of relations of authority. This plurality in production creates potential for significant instability in the relationship between practice and landscape. Political authority presents itself as the arena of spatial production of the last resort. Zoning laws, for example, reveal the ability of the political apparatus to intrude itself into the spatiality of other authority relations. However, this control over landscape production is never complete and often unsustainable. Hence, political landscapes are constantly shifting in response to factors far beyond immediate political relationships.

In attending to the production and reproduction of authority in space and over time, politics begin to emerge within a set of intertwined relationships. The chapters that follow address the spatial constitution of political authority within four distinct sets of historically conditioned relationships: the landscapes of geopolitics established in interpolity relationships; the territorial landscapes of the polity forged in ties between regimes and subjects; the settlement-centered landscapes of regimes forged within intra-elite ties and links to grassroots coalitions; and the architectural landscapes of institutions. In disarticulating each of these relationships, I do not mean to suggest that each is independent of the other. On the contrary, by treating them as components of a broad vision of the practical constitution of authority, the following chapters argue for a broadly encompassing account of political landscapes.

CHAPTER 3

Geopolitics

After the flood had swept over (the earth) (and) when kingship was
lowered (again) from heaven, kingship was (first) in Kish. . . .
Twenty-three kings (thus) ruled it for 24,510 years, 3 months, and 3½
days. Kish was defeated in battle, . . . its kingship was removed to
Eanna (sacred precinct of Uruk).

The Sumerian King List (from Oppenheim,
"Babylonian and Assyrian Historical Texts")

In *The Rise and Fall of the Great Powers* (1987), Paul Kennedy describes modern macropolitical history as a progress of successive "Great Powers" across the world stage. Such powers are born as they marshal economic and technological resources superior to their neighbors; they die, inevitably, as their commitments in the wider ecumene (particularly demands on the military apparatus) outstrip the available resources. Once proud polities fall into decline, the benefits and burdens of "greatness" are taken up by others, renewing the cycle. What endures over time, in Kennedy's account, is the temporal axis of political development—the organic pattern of ascendancy and collapse that slips intact across space. The geography of greatness, the distribution of international political authority in space that has been so thoroughly made and remade between the cardinal points of Kennedy's historical horizon (the sixteenth and twentieth centuries A.D.), holds little import for the constancy of the historical cycle itself. The rise of the Hapsburgs and the ascendancy of the United States can be understood as manifestations of the same transcendent macropolitical process—what John Agnew and Stuart Corbridge (1995: 19) refer to as "the apostolic succession of Great Powers"—precisely because the unique spatial configurations of geopolitical relationships in which each was enveloped are presumed to be of little significance to the production of the international order.

In its depiction of a macropolitical history based on a well-ordered cycle of ascendancy and collapse, *The Rise and Fall of the Great Powers* bears an uncanny resemblance to the Sumerian King List, a set of texts composed during the late third and early second millennia B.C. in southern Mesopotamia.[1] Like Kennedy's opus, the Sumerian King List presents the political history of Sumer and Akkad (the southern and northern portions, respectively, of southern Mesopotamia) as a highly regimented cycle in which one city's suzerainty over the region is followed inexorably by defeat and the ascendancy of a new place to geopolitical primacy.[2] The texts present a highly formulaic recitation of southern Mesopotamian politics from antediluvian mythical origins to historically documented rulers of the early second millennium B.C., tracing the movement of "kingship" from city to city and delineating the reigns of each dynast:

> Akshak was smitten with weapons; its kingship to Kish was carried. In Kish Puzur-Sîn, son of Ku(g)-Baba, became King and reigned 25 years; Ur-Zababa(k), son of Puzur-Sîn, reigned 400 years; Simu-dâr reigned 30 years; Ûsî-watar <, son of Simu-dâr, > reigned 7 years; Eshtar-muti reigned 11 years; Ishmê-Shamash reigned 11 years; Nannia, a stonecutter, reigned 7 years. 7 kings reigned its 491 years. Kish was smitten with weapons; its kingship to Uruk was carried. (Jacobsen 1939: 107–11)

Geopolitical relations—the geographic links and barriers between polities within an ecumene—are described not as spatial issues but rather as problems of history, of descent. Indeed, spatial relationships among contemporary polities are explicitly prohibited by the extension of sovereignty to all of Sumer and Akkad. The axis of the narrative rests not, as we might expect, on the tensions across borders or between capitals, but rather on the temporal order of succession.

1. Thorkild Jacobsen (1939: 140–41) argues that the extant versions of the Sumerian King List derive from a common progenitor, likely composed during the reign of Utu-hegal of Uruk in the late twenty-second century B.C. The most extensive versions of the texts extend through the reign of Damiqilishu of Isin, just prior to the ascendancy of Hammurabi to the throne of Babylon (Hallo and Simpson 1971: 38).

2. Sumer generally describes the area of southern Mesopotamia south of Nippur. The Akkadian heartland is generally described as the region north of Nippur to the lower Diyala river basin (see chapter 5).

It is in the recasting of the essentially spatial problems posed by the geopolitical organization of authority as problems of temporal succession that the Sumerian King List and *The Rise and Fall of the Great Powers* converge. The physical delineation of territory, the management of interpolity ties, and the production of new geopolitical alignments are ontologically redefined as problems of history's "emplotment" rather than the production of space. (On modes of historical emplotment, see White 1973: 29–31.) The two works share a conviction that the geopolitical landscape— the experiential, perceptual, and imaginative dimensions of interpolity spatial relations—is epiphenomenal to more deeply embedded historical regularities that shift "greatness" or "kingship" from place to place. Indeed, both works create their most profound impact in the definition of temporal patterns that operate independent of space and place.

A number of scholars have argued that the Sumerian King List was composed during the reign of the Isin kings in order to legitimate their rule as a lawful inheritance of the "natural" macropolitical order ordained by the heavens from time immemorial (Finkelstein 1979; Hallo and Simpson 1971: 88). Central to this ambition was a re-description of the privileged spatial position of Isin within macropolitical flows of commerce, tribute, and commands as the product of an ineluctable historical process rather than of the more contingent world of politics. Political relationships across space were thus sublimated to the regular consistency of a single model of the dynamics of the macropolitical order. For Kennedy, holding space as an absolute allowed for the singularity of macropolitical process that enabled him to predict (albeit with some caution) that the same organic progression of rise and fall that brought down the Austro-Hungarian empire would ultimately humble the United States. And it was on the strength of this prediction that the book sailed up the bestseller lists, articulating in metahistorical terms the then-prevailing American fears of national decline (an observation also made by Anthony Giddens in Giddens, Mann, and Wallerstein 1989: 328).

By drawing a parallel between the treatment of the geopolitical landscape in the two works, I do not mean to diminish the erudition of *The Rise and Fall of the Great Powers* nor the ingenuity of the Sumerian King List, but rather to highlight the insufficiency of descriptions of macropolitical order that arise from an absolutist ontology of space. Although they create compelling renderings of political history, both fail to provide— in the case of the Sumerian King List, perhaps intentionally occlude— an equally compelling account of the sources of geopolitical landscapes. We are thus left without an understanding of how authority on a macropolitical scale is actively produced in the spatial practices that assemble geopolitical landscapes. As a result, the landscape emerges only as a poorly defined stage on which "greatness" or "kingship" enter and exit. Due to this frustrating lack of spatiality, it is unclear what lesson for political practice might be derived from either the Sumerian King List or Kennedy's study other than surrender to the rhythms of history.

In this chapter, I examine political landscapes as they are produced within the practices of geopolitical relationships. A sustained debate over the geopolitical constitution of archaic polities has most recently been waged over the Classic Maya polities of the first millennium A.D., and so I will take these as my case study. However, let me first clarify in what sense I employ the term geopolitics. I do not mean to carve out a distinct scalar category delimited in space; rather, I want to highlight one set of practical relationships implicated in the constitution of authority among polities within an ecumene. As we shall see, geopolitical landscapes trip easily across numerous traditional scalar divisions, as profoundly embedded in a single architectural monument as in regional settlement patterns. Geopolitics here thus refers to the relationships of authority that are created within practices among polities across a given ecumene through the demarcation of difference, hegemony, exclusion, and inclusion. Geopolitical landscapes are produced and reproduced on the ground through physical barriers and borders, in evocative cues that signal relations of independence and obeisance, and in the imagination of the proper political order of the world.

Classic Maya Geopolitical Landscapes

The emergence of complex polities in the monsoon forests of lowland Mesoamerica, marked so indelibly by the massive stone temple pyramids at sites such as Tikal and Copan, stands as one of the more intriguing problems of Precolumbian political history (fig. 7). Over a century of archaeological research and an expanding corpus of epigraphic records (made possible by the decipherment of the Maya writing system in the 1960s) have provided an increasingly refined portrait of the foundations of Classic Maya polities (for an overview, see Lucero 1999). Glimmers of incipient political complexity in the lowlands during the Late Preclassic period (300 B.C.– A.D. 250; fig. 8) have been measured primarily in terms of surprisingly precocious programs of monumental construction. Foremost among these are the central precincts at Nakbe and El Mirador in the northern Peten. Whereas Nakbe is distinctive in its very early forays into large-scale construction—beginning as early as the Mamom phase of the Middle Preclassic (ca. 600–300 B.C.)—El Mirador is justifiably renowned for the extent and scale of its settlement core. Dwarfing even Tikal in size, the central building complexes at El Mirador all seem to have been constructed during the Late Preclassic, positioning the site at the center of any discussion of the emergence of Maya political complexity (Matheny 1987). Unfortunately, Nakbe and El Mirador have not hosted the same kind of sustained archaeological investigations to date as later Classic period sites such as Tikal and Copan, in part because of the difficulty in locating major research projects in what is today an exceedingly remote locale. Beyond the broad indications of political complexity embodied by big buildings, some Late Preclassic tombs also indicate the florescence of an elite differentiated from the general populace by radical disparities in the wealth that they carried with them to the grave. Burial 85 at Tikal, for example, held a single incomplete male skeleton (the head and thigh bones were missing). The body had been wrapped in cloth, and a greenstone mask with shell-inlaid eyes and teeth was sewn onto the bundle (perhaps to replace the missing head). In the grave were 26 ceramic vessels, a stingray spine (a symbol of self-sacrifice), and spondylus shell.[3]

The threshold between the Late Preclassic and Early Classic is generally defined by the appearance of carved stelae that mark the advent of the Maya hieroglyphic writing system. Maya inscriptions on these stelae (the primary medium that has survived from the Classic period) were primarily political writings, detailing elite interactions (such as royal visits), commemorating the accomplishments of rulers (often in warfare), and recording performances of sacro-political rites (such as blood-letting ceremonies). Contrary to the early-twentieth-century reconstructions of the Classic Maya as peaceful theocracies, it now seems clear that political authority in Maya polities was vested in large part in a hereditary elite who presided over a governmental apparatus sufficiently centralized to bring together the labor and materials necessary to produce the large settlement cores whose ruins we admire today. But what remains a matter of great debate is the geopolitical landscape produced by Classic Maya polities and the relations of political authority that this spatial order supported.

3. Ceramics from the grave were dated to the Chikanel phase of the Late Preclassic. Radiocarbon analysis of pinewood charcoal from one vessel yielded a date of A.D. 16 ± 131 (Coe 1999: 76).

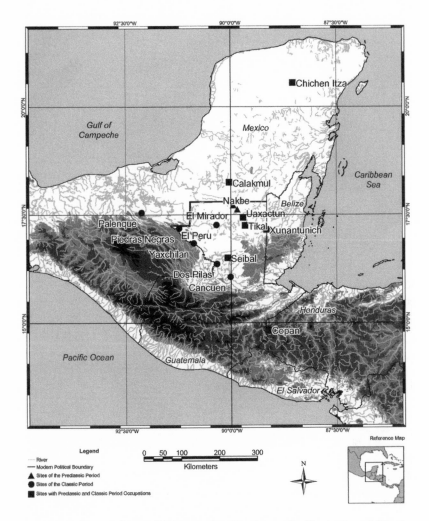

FIGURE 7. A map of central Mesoamerica showing major Preclassic and Classic period Maya sites. (Map source: ESRI Data & Map CD.)

The macropolitical order of the Classic lowland Maya has long been a focus of both epigraphic and archaeological studies. The problem has generally been invested in two competing visions of political organization that were delineated by J. Eric Thompson (1954: 77–81). Were Classic Maya polities organized as city-states "such as existed in Greece or medieval Italy, with political independence, but a fairly uniform culture and a common language," or were Maya communities integrated politically on a much

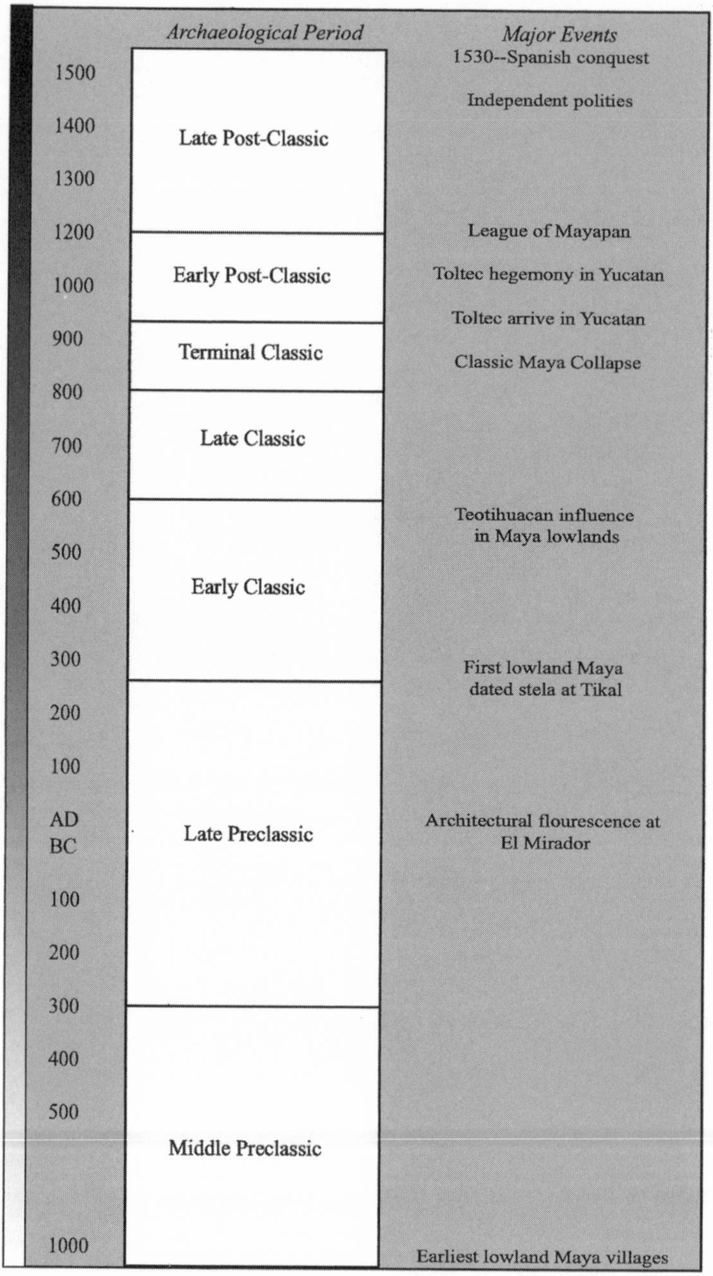

Archaeological Period	Major Events
	1530--Spanish conquest
Late Post-Classic (1500, 1400, 1300)	Independent polities
Early Post-Classic (1200, 1000)	League of Mayapan Toltec hegemony in Yucatan Toltec arrive in Yucatan
Terminal Classic (900)	Classic Maya Collapse
Late Classic (800, 700)	
Early Classic (600, 500, 400, 300)	Teotihuacan influence in Maya lowlands
Late Preclassic (200, 100, AD/BC, 100, 200)	First lowland Maya dated stela at Tikal Architectural flourescence at El Mirador
Middle Preclassic (300, 400, 500, 1000)	Earliest lowland Maya villages

FIGURE 8. Periodization and chronology of lowland Mesoamerica. (After Coe 1999.)

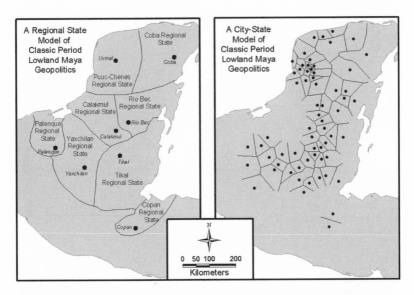

FIGURE 9. Two enduring models of Classic period lowland Maya geopolitical organization. (Redrawn from Mathews 1991.) **a.** The regional-state model. **b.** The city-state model.

larger regional scale with a small number of dominant states ruling large territories regulated by an integrated hierarchy of administrative loci? This city-state/regional-state dichotomy has dominated debates over Maya geopolitics to the present day, each position rising to prominence for a short time and subsequently declining in popularity. Sylvanus Morley (1946) originally proposed a model of a Maya regional-state, which he referred to as the "Old Empire," and early research at Tikal suggested to some that the site might have been *the* Classic period capital of the Maya lowlands (see the discussion in Marcus 1993: 113). Decipherment of Maya hieroglyphs and the expansion of large-scale archaeological projects to other major sites led to increasingly fragmented models of Classic Maya macropolitical order by raising the possibility that other settlements had rivaled Tikal as regional political capitals. In one of the more widely cited of these models, Richard Adams (1986) proposed an eight-polity configuration with large regional-states centered at Uxmal, Coba, Rio Bec, Calakmul, Yaxchilan, Palenque, Tikal, and Copan (fig. 9a).

The late 1980s and early 1990s brought a resurgence in the city-state view as it accorded with a group of popular "weak state" models that described Classic Maya polities as held together through the presumably more fragile bonds of kin or ritual performance rather than through the formalized rule of a centralized political apparatus (Ball and Taschek 1991; Demarest 1992; Chase and Chase 1996; Fox et al. 1996; Sanders 1989).[4] The emphasis on disintegration over integration promoted the further fracturing of the lowland geopolitical landscape into smaller autonomous political units, culminating in Peter Mathews's (1991) depiction of 63 discrete Classic Maya polities based on the lack of inter-site hierarchy that he read in the epigraphic record (fig. 9b). Recently, the pendulum has swung back to more regional views as the epigraphic record has revealed (*contra* Mathews) substantial evidence for hierarchy among rulers such that kings of large settlements such as Tikal and Copan were able to directly affect political decisions in neighboring settlements (Marcus 1993; Martin and Grube 1995, 2000; Coe 1999: 226–27).

Despite their manifest disagreement over the scale of Classic Maya polities, the city- and regional-state models are predicated on the same absolute spatial ontology. Both assume that the geopolitical landscape of the lowlands followed from the evolution of unitary types of political organization. Political forms are presumed to have adhered closely to a single formal type that was evenly distributed across the lowlands to create a homogeneous macropolitical order. These spatial forms do not originate in the practices of allied and rival polities but rather inhere in general types. Hence categorization of the lowland Maya as a homogeneous field of city-states carries with it a simple map of the regional macropolitical order.

A more complex spatial ontology underlies the "dynamic model" of Classic Maya political organization forwarded by Joyce Marcus (1993, 1998). Marcus rightly criticizes the city-state and regional-state models for their static vision of Classic Maya political history: "Data from the Classic Maya . . . support our notion of a dynamic system that changed over time, rather than exhibiting long-term stasis" (1993: 145). In their place, Marcus describes a repeating historical cycle of integration and disintegration derived from the political history of post-Classic Yucatán. Based on her analysis of sixteenth-century ethnohistoric sources, Marcus describes three general types of political organizations that characterized the Yucatán peninsula following the collapse of the Mayapán "empire": centralized polities governed by a territorial ruler residing in a capital city, a decentralized

4. The inspiration for the "weak state" models came from cultural anthropologists working in Africa (Aidan Southall's [1953] segmentary state) and southeast Asia (Stanley Tambiah's [1977] galactic polity and Clifford Geertz's [1980] theater state).

territorial polity ruled by a confederacy of local lords held together by kin ties, and a set of loosely organized groups of towns. Classic Maya polities, Marcus suggests, cycled through similar phases of organization:

> At the peak of each cycle, Maya states were territorially extensive and had a settlement hierarchy of at least four tiers, the upper three of which had administrative functions and were ruled by hereditary lords. At the low point of each cycle, formerly extensive states had broken down into loosely allied or semiautonomous provinces that sometimes had settlement hierarchies with only three tiers. (1998: 59–60)

Marcus embeds the ethnohistoric data in a traditional locational account of the State. This creates a peculiar epistemological ambiguity in the dynamic model, at once predicated on a presumed superior sensitivity of historical analogies derived from post-contact Yucatán rather than from the Old World, yet eschewing the very specificity of the ethnohistoric case in order to force the data into rigid universal categories of settlement hierarchy.[5] This absolute spatial sense rests not only on her reiteration of Gregory Johnson's assumption that the form of the macropolitical landscape will follow unproblematically from the structural type of the polity (see chapter 1) but also on an informally posited stability in the geopolitical landscape (contrary to the "dynamic" label attached to the model). Marcus posits, albeit somewhat gingerly, that Maya polities coalesce and decompose along a regular set of "cleavage planes," suggesting an enduring spatial order of political integration and disintegration independent of actual political practices.[6] As a result, even though

5. Marcus's argument for the priority of models drawn from the immediate region is undercut by her veneration for the universal, dismissing historical description that does not derive from "nomothetic regularity" as appealing to "tropical mysteries" (1993: 115). If this were indeed the case, then there would be no de facto reason to grant priority to a model drawn from the Yucatán over one drawn from anywhere else (an assertion that Marcus defends quite well). This tension between the universal and particular creates a degree of epistemological confusion because the central-place models that are ceded analytic priority over the ethnohistoric records were adopted from geographic studies of Germany (Christaller and Baskin 1966) and from an archaeological study by Ian Hodder (1972; Hodder and Hassall 1971) of Roman Britain—pretty far afield from the Mesoamerican lowlands. Norman Hammond's original reasoning for importing locational methods to the study of the Classic Maya directly contradicts Marcus's preference for models drawn from local political history. Hodder's success in using Christaller's methods to describe settlement in Roman Britain, Hammond (1974: 315) argues, "shows that the model can and should be applied to the Classic Maya situation."

6. "What would be interesting to know is whether the sixteen provinces left after the collapse of Mayapán were the same units that had come together to form the *mul tepal*. If

she provides a well-developed argument for embedding models of Classic Maya geopolitics within a specifically Mesoamerican political history (a thoroughly historicist privileging of ethnohistory over metahistory), Marcus's model ultimately reiterates an absolute spatial ontology by linking the spatial forms of polities to a regular developmental sequence rooted in social evolution rather than an account of geopolitical practices.

Like the models of political transformation offered by *The Rise and Fall of the Great Powers* and the Sumerian King List, the central premise of Marcus's analysis is that the geopolitical organization of the lowlands was organized by a regular temporal cycle of emergence and collapse, rise and fall. The spatial organization of the lowland geopolitical landscape follows from the evolution of political forms, but in no way did these spaces direct or influence the future transformation of political relationships or the bases of authority. If the landscape were allowed to hold implications for the constitution of geopolitical orders, then that would require an account of the political production of space, not just a typology of political forms. Although a number of Mayanists have led the way in redescribing the built environments of Maya settlements as instruments of political authority, the spatial order of the geopolitical landscape remains hemmed in by a firmly ensconced absolutism. However, there are some indications that the study of Classic Maya geopolitics is moving away from such deeply sublimated descriptions of space and toward an account of the production of authority in the experience, perception, and imagination of landscapes.

EXPERIENCE

Since the 1970s, archaeological approaches to the physical ordering of the Classic Maya geopolitical landscape have relied almost exclusively on the procedures of locational geography, applying the methods of central-place analysis to Maya settlement distribution. These studies have been most attentive to the formal dimensions of landscape, examining geometric relationships between sites with an eye to describing regular patterns in settlement location. In one of the earliest such studies, Norman Hammond (1972) enveloped the central area of the lowlands in a geometric settlement lattice, representing the territorial extents of polities in terms of Thiessen polygons drawn around major sites. Boundaries between hy-

---

so, the boundaries of these provinces might have been the 'cleavage planes' along which the regional state was likely to break up when the time came" (Marcus 1993: 121).

pothetical territories were defined by halving the distance to nearby settlements. (The polygon technique for describing settlement locations is
described at some length in Lösch 1954: 116–30.) Because of the irregular density of Maya site distributions, the result was a map of the lowlands that parceled the area into a honeycomb of variously sized territories (fig. 10a). In a coeval study, Kent Flannery (1972: 420–23) constructed
a nearest neighbor settlement lattice for the northeast Peten, connecting
each site in a straight line with every adjacent site. The resulting point
and line diagram described absolute distances between major sites that,
in contrast to the uneven spaces of the Thiessen plot, emphasized the geometric regularity of site distribution (fig. 10b). The goal of this exercise
was to define the fundamental settlement geometry that results from the
posited evolution of complex systems (social and environmental) and the
universal rules that order the rise of the State.

These studies were of considerable import to the examination of Classic Maya geopolitics, less for the actual models that they built for the political order of the lowlands than for the fundamental assumptions within
which they framed the problem and its archaeological solution. It should
be noted that both Hammond and Flannery were skeptical of the potential
for the central-place methods that they employed to discern actual political territories. For Hammond (1974: 322), these analytical procedures
served as starting points for guiding field investigations of political
boundaries. For Flannery, the simple diagrams of locational geography
(which he later derided as demanding "all the rigor of those follow-the-
dots puzzles you used to work out as a kid" [1977: 661]) were simply a
means for exposing the essential geometry underlying complex social formations produced by the rules ordering the cultural evolution of civilizations and the archaeological record.[7] But the theoretical consequence
of these studies was to embed investigations of Classic Maya politics
within a set of methodological procedures that equate the investigation
of geopolitical relations with the discovery of universal principles ordering macropolitical space. By lashing Classic Maya geopolitics to an absolute ontology of space in which the spatial form of the macropolitical
order followed directly from the typological classification of sociopolitical
structure (which was itself ultimately given shape by the social evolu-

7. "The ultimate goal of a systems analysis might well be the establishment of a series
of rules by which the origins of some complex systems could be simulated. . . . Let us, therefore, conclude by tentatively putting forth fifteen rules out of the scores with which we might
one day be able to simulate the rise of the state" (Flannery 1972: 421).

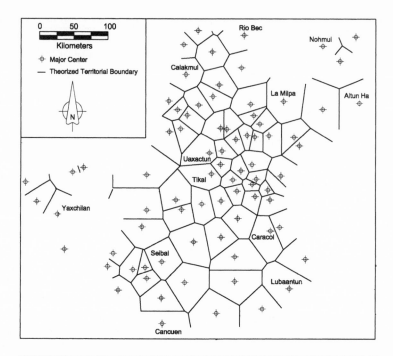

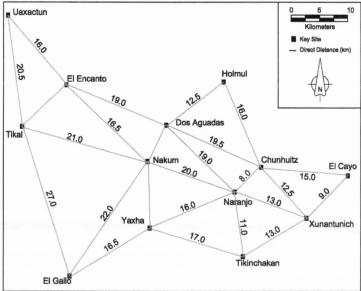

FIGURE 10. **a.** A Thiessen polygon lattice applied to Classic Maya settlement. (Redrawn from Hammond 1972). **b.** A nearest neighbor array—Classic period Maya, Peten region, Guatemala. (Redrawn after Flannery 1972.)

tionary process), the development of a sense of the active production of geopolitical orders within political practices was explicitly forestalled.

Epigraphic investigations of the Classic Maya geopolitical order were no less shaped by the absolutist commitments of archaeologists. One epigraphic approach to the problem of Classic Maya geopolitics has emerged from Heinrich Berlin's 1958 observation that a unique set of glyphs on stelae distributed at sites across the lowlands shared certain lexical and syntactical features even as the main signs of each had a very limited geographical distribution, usually confined to a single site. Berlin termed these signs "Emblem Glyphs," suggesting that they represented in some way the settlement in which they occurred. Marcus expanded the toponymic reference of the Emblem Glyphs, arguing that they referred not just to a single city but to the entire political territory subject to that city. Although the territorial referent of Emblem Glyphs is now generally agreed on, the geopolitical landscape that follows from the information contained in the inscriptions remains the subject of debate. Mathews, a staunch defender of the city-state position in recent debates, has suggested that each site boasting an Emblem Glyph on its stelae constituted a politically autonomous unit controlling an independent territory (1991). The presence of an Emblem Glyph at a site, he argues, denotes the presence of Maya rulers of equally elevated rank. The effect of this reconstruction is to fracture the geopolitical landscape into 63 independent polities (see fig. 9b). Not only does Mathews's map of the Maya region closely resemble Hammond's honeycomb Thiessen plot, but it is also predicated on a similar willingness to read a singular spatial form as following directly from a singular political type.

Archaeological study of the experiential dimensions of the Classic Maya geopolitical landscape was thus reduced to a search for regular patterns in settlement location that might serve as proxy measures of organizational form, particularly the spacing of centers and the size of "sustaining areas" around major sites. This search was organized by two key assumptions regarding the physical organization of polities: coherence and continuity. The use of polygons to describe political territories of Maya polities was predicated on the assumption that the macropolitical order of the lowlands was continuous; that is, where one polity left off, another began such that every square meter of land was allotted to the control of one polity (an assumption that Hammond [1974: 322] carefully noted to be highly doubtful). The primary bulwark to this assumption has been the observation of uniform spacing between settlements that Flannery (1972: 421) interpreted as an indication of a high degree of integration between

Maya "centers."[8] In the Belize valley, Classic centers have been described as approximately 10–15 kilometers apart (Willey et al. 1965: 573); in southern Quintana Roo, Peter Harrison (1981: 274–76) suggested that major centers were placed at intervals of 26 kilometers whereas minor centers appeared at 13-kilometer intervals; and in the northeast Peten, Marcus (1973) found the average interval to be 15.8 kilometers. This regularity in spacing revealed through the application of central-place methods is presumed to reflect the underlying administrative regularity of Classic Maya polities themselves.

Marcus captured the logic of this remarkably durable argument well in positing that a close match between the distribution of Classic centers and *"an ideal hexagonal lattice would not be present if the assumptions of administrative hierarchy were false"* (Marcus 1993: 154, italics in original). This argument places the burden of proof on negating the claims of central place analyses to explanatory exclusivity (rather than systematically ruling out alternatives to bolster such arguments), tightly restricting the imagination of possible determinants of spatial form. More seriously, it also allows the results of analysis (an even lattice of hierarchically integrated centers) to confirm the starting assumptions (polities organized into administrative hierarchies)—a rather circular logic.

But were major Classic Maya centers as uniform in their distribution as they are often described? If we follow Flannery by taking the major sites of the northeast Peten region between Uaxactun and El Cayo as our case study, we can examine the variation in distances between sites and their neighbors (table 1). Utilizing the nearest-neighbor distance data for 13 sites in the region, we arrive at a mean site distance of 11.54 kilometers.[9] The standard deviation within the data set is 3.88, resulting in a rather high coefficient of variance of 33.59. If we carry out the analysis to the second- and third-nearest neighbors, the results are no more regular. The coefficient of variance for the second- and third-nearest neighbor data sets are 23.58 and 43.68, respectively, much too high to plausibly suggest that locational decisions were structured by considerations of direct distance

8. The term "centers" refers to "aggregates and nucleated arrangements of pyramids, big platforms, palaces, and other buildings that were the foci of Maya political and religious life" (Willey 1981: 391). Gordon Willey goes on to note that the lack of a modifying adjective (such as "political" or "religious" center) derives from a desire both to highlight the likely multifunctionality of sites and to remain noncommittal about the fundamental character of Maya urban spaces.

9. In this discussion I use the distance data published by Hammond (1972). Flannery's (1972) published distances were rounded and thus less precise.

TABLE I. Nearest Neighbor Distances for Classic Period Maya Sites
in the Northeast Peten

Site Name	Nearest Neighbor Distances (km)		
	First	Second	Third
Uaxactun	13.5	16.5	19.5
Tikal	12	19.5	21
El Encanto	12	14.5	16.5
Holmul	11.5	13.5	18.5
Dos Aguadas	11.5	11.5	15.5
Nakum	11	11.5	17
Chunhuitz	6	13.5	14.5
Naranjo	6	10	14
Yaxha	11	15.5	17
Tikinchakan	10	14	15
Xunantunich	9	12.5	13
El Gallo	15.5	17.5	21.5
El Cayo	21	22.5	45.5

to neighboring settlements.[10] Such a significant degree of variation within the data casts substantial doubt on the description of Classic Maya centers of the northeast Peten as geometrically uniform in their spacing and on the presumption of a continuous geopolitical landscape.

Furthermore, it is worth questioning the utility of straight-line distance measures used in locational settlement lattices that express the relationship between sites "as the crow flies." What meaning do such distances actually hold for real networks of transport and communication? A report on the results of a remote sensing project conducted in the region around Calakmul claims to have verified the radial lattice of roads linking sites demanded by central-place theory (fig. 11; Folan, Marcus, and Miller 1995). The authors describe the distribution of the detected Maya raised roads, or *sacbeob,* as showing "a good fit" with the idealized lines connecting Calakmul to its dependencies demanded by central place theory (ibid., 277). Although the data provided on the network of linked pathways are quite remarkable and show the promise of remote sensing techniques as applied to archaeological problems, the results do not seem to provide the kind of archaeological confirmation of the central-place

10. These are different statistical results than those arrived at by Flannery (1972: 421) and Hammond (1974: 322–26).

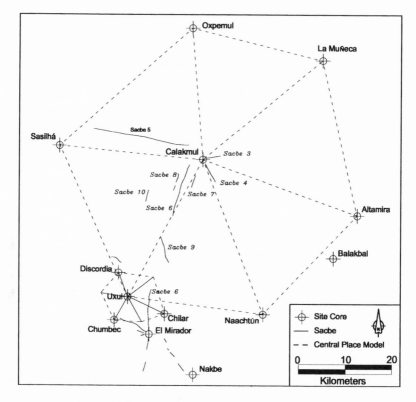

FIGURE 11. A central place hub-and-spoke array superimposed on remotely sensed *sacbeob* around Calakmul. (Redrawn after Folan, Marcus, and Miller 1995.)

model that the authors suggest. First, a potential fit only exists for three of the six radial lines—sacbeob that depart from Calakmul pointing in the general direction of Naachtun, Sasilha, and either Uxul or El Mirador. Of these, Sacbe 4 is simply too short to hypothesize its ultimate endpoint (only approximately 5 kilometers of the 30-kilometer distance between Calakmul and Naachtun was detected). Furthermore, Sacbe 6, which the authors link to El Mirador, does not link Calakmul to a dependency but rather to a Preclassic rival (El Mirador appears to have been largely abandoned by the Classic period). This leaves only Sacbe 5 as providing a clear potential pathway between Calakmul and a Classic period dependency, suggesting that the fit between the actual sacbeob discovered through remote sensing and the lattice of central-place theory is actually quite poor.

The central issue is not the tightness of fit between lines on the ground

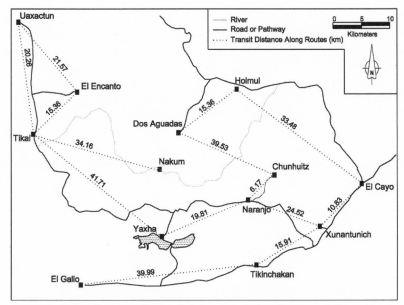

FIGURE 12. Site distances along existing routes and pathways, northeast Peten, Guatemala. (Source: author.)

and central-place lattices, but whether such lattices provide an appropriate way to model relationships between places within a geopolitical landscape. It is certainly of less significance, in terms of spatial experience, that a straight line between Chunhuitz and El Cayo, for example, is 15 kilometers than that the distance along the roads that link them is approximately 35 kilometers. Figure 12 describes the physical links between Classic period centers of the northeast Peten by using modern trails, roads, and river routes to connect them.[11] This representation of the interconnections between sites departs significantly from both the hub-and-spoke

11. I do not presume that modern routes mimic ones used during the Classic period but merely suggest that actual movements across landscapes are of considerably greater import to an understanding of spatial relationships than idealized connect-the-dot drawings that produce regular lattices. There are many unique factors that have contributed to the layout of modern pathways—including, of course, the tourism generated by major archaeological sites. Thus, there is great danger in creating analogies between modern and ancient pathways. Nevertheless, modern routes do provide a useful check upon the idealized lattices of locational analyses, forcing a reexamination of core assumptions about the factors forging settlement patterns. A second limitation of the settlement pathway network described in figure 12 arises from the manner of their discovery. William Bullard's (1960) survey of the northeast Peten followed available trails through the region, documenting the sites along

system of the idealized locational lattices and the linear beaded model suggested by a recent study of settlement in the La Entrada region of Honduras (Inomata and Aoyama 1996). Two different relationships between sites and pathways emerge from this depiction—sites located directly on a primary circuit (such as Uaxactun, Tikal, and El Cayo) and sites set into cul-de-sacs (such as El Encanto, Dos Aguadas, and Chunhuitz). Such an irregular pattern of linkages suggests that the Classic Maya geopolitical landscape may not have been as continuous as it is usually described, with variably isolated pockets of settlement irrupting where the controlling political authority was less than perfectly clear. In such a case, it would seem better to look for frontiers between polities, zones of steadily decreasing control exerted by a central governmental apparatus negotiated within geopolitical practices rather than simple boundaries linking polities into a geometric mosaic. But such an examination would require that we attend less to the evolution of political forms and more to the constitution of authority relations between polities as they are negotiated on the fringes of territory. (For more highly elaborated arguments describing the practices that establish frontiers and boundaries, see Agnew 1999 and Boone 1998.)

In a recent study of Classic period warfare at La Pasadita, Charles Golden (2003) has persuasively argued that the political frontiers in the Usumacinta river basin showed a remarkable fluidity over time. Although the area of La Pasadita had remained largely interstitial for much of the Classic period, as competition between Piedras Negras and Yaxchilan intensified in the eighth century, the rulers of the latter attended more closely to activities at this well-defended frontier site. By enveloping the elites of La Pasadita within the patronage of Yaxchilan's king, Bird Jaguar IV, this previously unincorporated space became part of an increasing crystallization of frontiers between competing polities. Golden's study reveals the political practices of patronage and competition that produced La Pasadita as a frontier within the shifting political landscape in order to securely constitute the expanding authority of the Yaxchilan regime. Golden's study thus not only undermines the assumption of continuity that has traditionally guided models of Maya geopolitics but also strongly roots an understanding of geopolitics within shifting multiple relationships (such as the triangulation between Yaxchilan, Piedras Ne-

---

them. Thus the intersection of sites and trails is not coincidental; it was built into the research design. Nevertheless, it is instructive to replace the connect-the-dots diagrams of central-place theory with maps that better reflect the problems and potentialities for regulating flows between sites.

gras, and La Pasadita) and the practical production of transformed political landscapes.

A second assumption underlying locational descriptions of political
domains is the implicit contention that Classic Maya political territories
were coherent, that sectors of authority—what has unfortunately been
glossed in economic, rather than political, terms as the "sustaining area"
of a center—clustered closely around central places. It is this assumption
that has raised consistent protestations against the results of the Thiessen
polygon method of defining territories. The honeycomb map (see fig.
10a) of the lowland geopolitical landscape that this method creates leaves
the large site of Tikal (approx. 1,600 hectares) stranded on a territory
roughly the same size as that accorded the small center at El Encanto (a
site less than 1 hectare in size). The problem lies with the appropriation
of locational models that describe ideal *economic* locational geometries to
model real *political* territories. The method and theory of central-place
approaches were developed to describe economic integration, to model
spatial configurations that emerge within market-based economies under idealized topographic conditions, such as those presented by the open
plains of southern Germany (Christaller and Baskin 1966).[12] Application
of these analytical methods to describe political integration requires that
we assume a priori the complete coherence of a political domain, with
no allowance made for sectors of control that might lie in noncontiguous spaces (such as describes the lands under the control of the Palestinian Authority or, for that matter, the territory administered by the
United States). The problem is that, whereas assumptions of "least cost"
behavioral motivations may prove relatively safe within a politically
unified, market-driven economic system, they generally prove tautological for analyses of nonmarket economies whose political integration is
the very question at hand.

A recent study of verbs related to political action contained in the epigraphic record casts substantial doubt on the coherence of Classic Maya

12. The distinction between real settlement distributions and economically ideal locational decisions was explicitly noted by August Lösch (1954), who distinguished the "question of actual location . . . from that of the rational location." Lösch and most locational
geographers were specifically interested in the latter because they aspired to develop models of how locational choices should be made, not how they have been made in the past.
Archaeologists who have employed the methods of locational geography have not fully
appreciated that these techniques were developed to be proscriptive of an ideal rational
settlement distribution under conditions of market capitalism, not necessarily descriptive
accounts of all human settlement.

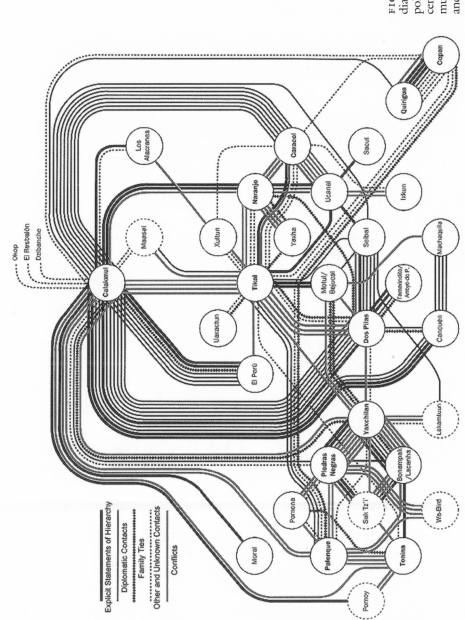

FIGURE 13. A diagram of bilateral political relationships centered on Calakmul. (After Martin and Grube 2000.)

polities. Simon Martin and Nikolai Grube (1995, 2000) have reconstructed discrete sets of bilateral relationships between the large center at Calakmul and nine of its peers (fig. 13). The geographic implications of their findings bear directly on the issue of the spatial continuity of Classic Maya polities. The inscriptions that Martin and Grube examined detail Calakmul's repeated armed conflicts with Tikal and Palenque, clearly establishing these polities as autonomous rivals and setting the broad spatial parameters for the political territory under Calakmul's immediate authority. The texts also detail diplomatic exchanges between Calakmul and allied polities. Gift giving among rulers, jointly conducted rituals, and marriage exchanges linked Calakmul to a web of allies who appear to have coexisted as geopolitical peers, such as Caracol and Yaxchilan.

In addition to enemies and friends, the inscriptions record subordinate polities over whom Calakmul exerted considerable authority. In a number of contexts, what Martin and Grube describe as "overlord statements" record the accession of kings as having taken place under the aegis of the superior king of Calakmul. For example, a panel from Cancuén refers to two episodes of kingly accession (in A.D. 656 and 677), described as having occurred under the auspices of, or perhaps more accurately through the direct agency of, the king of Calakmul (Martin and Grube 1995: 44). In addition to Calakmul's supervision of the accession of local kings, Martin and Grube describe explicit statements of the subordination of local rulers to the king of Calakmul, such as that found on the hieroglyphic stairway at Dos Pilas. These statements provide explicit records of geopolitical relationships of authority in which Calakmul exercises a degree of control over noncontiguous polities, casting doubt on descriptions of Maya polities as spatially coherent and enveloping these sites in a much more complex landscape of geopolitical authority than can be modeled with simple polygons.

Martin and Grube's analysis necessitates a revised cartography of the experience of the Classic period geopolitical landscape during the seventh century A.D. (fig. 14). Based on accounts of interpolity hostilities, we can posit three major rivals to hegemony (but not suzerainty) in the central area of the lowlands: Calakmul, Palenque, and Tikal. Other polities, such as Yaxchilan and Caracol, were treated as allies and thus also must be counted as peers, even if their spheres of influence were more restricted. The inscriptions also describe polities subordinate to Calakmul that, in terms of geopolitics, must be treated as satellites though probably not dependencies: El Peru, Naranjo, Cancuén, Dos Pilas, and possibly Piedras Negras. Although both El Peru and Piedras Negras are somewhat far afield

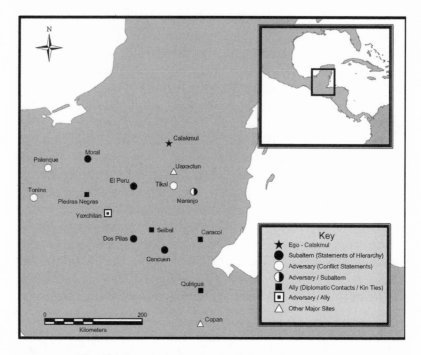

FIGURE 14. The Classic period Maya lowlands from the perspective of Calakmul: a cartography of rivals, dependents, and allies. (Map source: ESRI Data & Map CD.)

from Calakmul—more proximal to Tikal and Palenque—it is possible to suggest a system of frontiers in which these sites remain interior to a continuous Calakmul territory. This is a more problematic exercise for Naranjo, which lies about 60 kilometers southeast of Tikal. However, it is difficult to conceive of Dos Pilas or Cancuén as components within a continuous Calakmul polity. If these sites are to be considered as falling under Calakmul's authority in a geopolitical sense during part of the Classic period, then we must dispense with the presumption of continuous political territories and begin to describe, as Martin and Grube have done, the practical relationships between polities.

What is most important about Martin and Grube's analysis to an account of the political landscape is that they provide a description of the practical means by which the experiential dimension of geopolitical space was actively produced within varying relationships of authority among polities. Through warfare, marital alliances, and diplomatic exchange, the experiential dimension of the geopolitical landscape was produced and

reformed as coalitions were assembled and recalcitrant allies were sub-
dued. Tikal appears to have been Calakmul's central strategic preoccupa-
tion. Taken as a synchronic view, the map of the geopolitical landscape
presented in figure 14 is most remarkable with respect to the encirclement
of Tikal by polities either vassals of, or allied with, Calakmul. The geopolitical view from Tikal in the late seventh century A.D. must have been
quite dire, with formidable adversaries arrayed all around it linked through
the ruler of Calakmul. However, Tikal appears to have escaped the tight-
ening noose when, in A.D. 695, it defeated Calakmul and (probably) captured its king, Yich'aak K'ak' (a.k.a. Fiery Paw; Martin and Grube 2000:
44–45). Tikal subsequently conducted successful wars in the eighth century against El Peru (A.D. 743) and Naranjo (A.D. 744), effectively ending
the geopolitical coalition that Calakmul had assembled against it.

What Martin and Grube have found in their analysis of the Maya hieroglyphs are the practical, historically rooted, interpolity relationships that
produce geopolitical landscapes. Where archaeologists employing central-
place theory and epigraphers seeking singular principles of political organization had sought nomothetic regularities—general rules of politics
as elaborated in space—Martin and Grube have found variation, contestation, and shifting political relationships predicated on changing landscapes. Under the weight of the details recounted in the historical records,
generalizations about *the* Maya geopolitical landscape seem grossly inadequate to expressing the dynamics of political expansions and collapses.
What we end up with, then, is a varied geopolitical landscape created within
the practices of rival polities (conquest, intimidation, accommodation)
rather than a lattice generated by the absolute nature of political spaces.

Although Martin and Grube's reconstruction provides a foundation
for rethinking the experiential dimensions of the Classic Maya geopolitical landscape—the physical relationships between places that constituted
varying relations of authority—it also begs the question as to how such
relations were marked in the perceptual dimension of landscape. How
did the form of the landscape carry signals regarding shifting geopolitical relations of authority?

PERCEPTION

Production of political landscapes hinges not simply on the production
and enforcement of relations of authority and subjection as experience—
that is, in the movement of people and things across physical space—but
also on the fostering of an enduring perception of geopolitical relation-

ships. How was such an understanding of relations of authority and subjection produced and reproduced in the landscapes of Maya polities? Three sets of political practices stand out as creating an instrumental built aesthetic through which geopolitical relationships were reified: memorialization, emulation, and authorization.

Memorialization is the most overt mode of rendering geopolitical relations because it encompasses features whose aesthetics are explicitly directed toward cueing memories of specific events that define a polity's role within the macropolitical order. Memorialization can thus be a medium for boasting of a polity's superiority or reinforcing another's subjugation. The hieroglyphic stairway at Seibal provides a good example of the latter genre because it inscribed in the architecture of the polity's center its subjugation within the geopolitical landscape. In A.D. 735, Dos Pilas defeated and captured the king of Seibal, erecting a hieroglyphic stairway in the central plaza to remind the vanquished of their subjugation (Schele and Mathews 1998: 177). Memorialization also took more triumphal forms, providing testimony to the geopolitical superiority of a polity. When Hasaw-Kan-K'awil, ruler of subjugated Tikal, reasserted Tikal's independence in A.D. 695 by defeating and capturing the king of Calakmul, he erected a stucco frieze depicting the event that "dominated the space where he conducted the business of the court" (ibid., 86).

Victorious Dos Pilas also claimed, in stelae erected at both Dos Pilas and Aguateca, to have also demolished Seibal's monuments, to have "destroyed the writing" (ibid., 177). The absence of monuments at Seibal detailing political history prior to subjection by Dos Pilas may be a consequence of this practice. Such concern to re-describe the geopolitical landscape as well as blot out previous accounts suggests that memorials are not simply passive expressions of political power but rather play an instrumental role in reproducing authority in specific formations. Similarly, at Yaxchilan pieces of old monuments were set into new compositions, suggesting active moves by rulers to rearrange the relationship between past and present (Tate 1992).[13] Memorials thus work within larger geopolitical landscapes not simply by recording political facts of domination and subjection but also by creating and reproducing that fact on the ground by reinforcing the perception of a polity's position within the complex relationships of the geopolitical landscape.

Emulation describes a considerably more subtle perceptual dimension

13. Thanks to Cynthia Robin for pointing out to me the rearranging of Yaxchilan monuments.

of geopolitical landscapes because it embraces built aesthetics with less direct political referents. Wendy Ashmore has noted, for example, how the cores of Naranjo and Xunantunich hold a remarkable formal resemblance to the central sector of Calakmul: "To me, such pronounced similarity implies deliberate emulation by the later, smaller sites [Xunantunich and Naranjo]—and perhaps by others, yet unrecognized" (1996: 9). She goes on to suggest a reason for such emulation: "Given the importance increasingly attributed to Calakmul's role in pan-lowland Maya politics, over many centuries . . . it may well be that the developing form of that city provided templates for copying at multiple younger centers at different points in time" (ibid.).

Xunantunich's emulation of Calakmul may extend beyond settlement layout to include the architecture of major structures. Ashmore suggests that one such structure (A-6) "alludes directly to the relative height and form of Calakmul's structure II" (ibid.). The allusion of structure A-6's form would have complemented the iconography of the adorning stucco frieze that detailed themes of dynastic succession and the religious foundations of political authority. Ultimately, Ashmore suggests that the impetus to emulate Calakmul lies in the shifting geopolitical landscape of the central lowlands at the time of Xunantunich's founding in the seventh or eighth century A.D. Indeed, the emulation of Calakmul, rather than the more proximal hegemonic power at Tikal, may have constituted as much a rejection of the latter's pretensions to local primacy as an affirmation of Calakmul's authority. Although there is nothing to suggest that Xunantunich was ever subject to Calakmul, the very nature of a hegemonic geopolitical power is vested as much in prestige as in actual domination. Hence the emulation of Calakmul bolsters not only the authority of local rulers within the polity but also the geopolitical authority of Calakmul within the larger ecumene.

In contrast to memorialization and emulation, authorization describes an aesthetic expression of legitimate empowerment whereby a polity expresses its status as an important feature of the geopolitical landscape. Unlike practices of memorialization that focus on representations of specific events, the aesthetics of authorization center on much broader rhetorical politics that position the polity within a symbolics of the natural or cosmic order. A great deal of provocative research has recently been accomplished on Classic Maya built aesthetics that would fall under the rubric of authorization. Ashmore's 1991 study of Maya site-planning principles paved the way for archaeological analyses of the perception of landscapes generally and the political production of cosmographic cities more

FIGURE 15. The House of the Governor, Uxmal. (Photo courtesy of Patricia Cook.)

specifically (see also Brady and Ashmore 1999). In one particularly nuanced study of built aesthetics, Jeff Kowalski and Nicholas Dunning (1999) examined the Terminal Classic (ca. A.D. 770–950) site of Uxmal, documenting the use of architectural elements to signal regional political suzerainty. Several of the monumental buildings at the site seem to clearly argue for the supremacy of Uxmal within a regional political sphere. The upper facade of the House of the Governor, a 160 × 133 meter stone masonry construction set atop an elevated platform 7.4–11.8 meters high, includes a stone mosaic shaped into an interweaving lattice pattern and a step-fret design (fig. 15). The former pattern, Kowalski and Dunning note, duplicates the warp and woof of a woven mat, an enduring Maya symbol of rulership, whereas the latter designated elites within northern Maya political symbologies as well as those of Oaxaca and the Gulf Coast region. The eastern facade of the building is even more explicit in its rhetoric. Here, sculpted figures of persons of various political ranks culminate in the depiction of a supreme individual, whom they interpret as Lord Chaac, ruler of Uxmal, over the central doorway.

A second monumental structure at Uxmal, built during the reign of Lord Chaac, is the Nunnery Quadrangle. Kowalski and Dunning argue that this structure, oriented to the cardinal directions, served as a built representation of the quadripartite division of the cosmos within Maya mythology, elaborating the multilayered conceptions of the Upperworld (thirteen layers), Middleworld (seven layers), and Underworld (nine layers) into the built form of the North (thirteen entryways), West (seven entryways), and South (nine entryways) structures, respectively.[14] At the center of the Quadrangle's courtyard, an upright stone column echoed the "First Tree of the World" as an axis mundi of the cosmos. Near the remains of this column was found a single-headed jaguar throne echoing a Maya connection between the king and the first tree of the world, thus associating the political authority of the king with the "primordial acts of world creation": "In the Nunnery Quadrangle, 'Lord Chaac' and his architect seem to have made a conscious effort to embody key elements of essential Maya cosmological concepts in the plan and sculpture to convey the idea that Uxmal had become the religious center and political capital of the eastern Puuc region" (Kowalski and Dunning 1999: 287, 279–80). At issue is not simply a model of the cosmos but also an argument for the central position of Uxmal within that cosmos.

The perceptual dimensions of the geopolitical landscape describe the relationships of domination and subjection among polities so as to reproduce these political relationships on the ground—to reinforce sensibilities of defeat or triumph. This is not simply an embedding of history within the built environment but also an attempt to use that environment as an instrument in realizing political goals. Yet what remains to be explored is how both the experiential and the perceptual dimensions of the geopolitical landscape were articulated within a shifting cultural sense of the proper spatial order of Maya polities. Such an account demands that we turn to the imagination of geopolitical landscapes.

IMAGINATION

In A.D. 732, Waxaklahun-Ubah-K'awil (a.k.a. 18 Rabbit or 18 Jog), ruler of the Maya polity centered at Copan, erected a stela (known as Stela A) in that site's Great Plaza (fig. 16). The front of the stela shows the ruler in frontal perspective dressed in ornate regalia, perhaps imitative of Kan-Te-Ahaw, patron deity of Copan (Schele and Matthews 1998: 158). On both sides of the stela, an inscription records the date of the monument's erection and the rituals attendant to its dedication. The final passage of the inscription is particularly notable because it has been interpreted as outlining a quadripartite model of Maya geopolitics, an "emic view" (Marcus 1993: 150) of the Maya political world as seen by the ruler of one of

14. The East and West together represent the Middleworld as the directions of the rising and setting sun. Interpretation of the East structure at the Nunnery Quadrangle, a structure atop a platform the same height as the West structure punctuated by five doorways, is less certain.

FIGURE 16. Copan Stela A. (From Maudsley 1889–1902: plate 25.)

the Maya lowland's more enduring polities. As deciphered by Linda Schele and Peter Mathews, the final passage of Stela A reads:

North (xaman) gourd tree
Hao Ha four te skies
Four na skies, four ni skies
Four "deerhoof" skies, Holy Copan Lord
Holy Tikal Lord, Holy Kala'mul Lord
Holy Palenque Lord, he did something
??? sky, ??? earth
Lak'in (east), ochk'in (west)
Nohol (south), xaman (north)
Hao Ha, it was opened
The ??? hole, it was closed
The ??? hole, at the middle of
???, ??? (1998: 160)

A similar stela (Stela 10) was recovered at the site of Seibal (fig. 17). Erected in A.D. 849, the front of the stela depicts Wat'ul, ruler of Seibal, wearing the same headdress as that donned by Waxaklahun-Ubah-K'awil on Stela A at Copan. The final passage of the adjacent inscription reads:

Holy Lord of Seibal
They witnessed it
Hun . . . K'awil/Holy Lord of Mutul (Tikal)
[and] Kan-Pet/Holy Lord of Kan (Kalak'mul)
[and] Kan-Ek'/Holy Lord of Nal (Motul de San José).
It happened in/the center of Seibal[.] (ibid., 186)

Instead of a quadripartite world centered on Copan, Tikal, Palenque, and Calakmul, Stela 10 at Seibal gives us four different pillars of lowland macropolitical order with Seibal and Motul de San José substituted for Copan and Palenque.

In a deservedly influential reading of the two stelae, Marcus interprets these inscriptions as descriptions of the macropolitical order of the Classic period lowland Maya rendered in cosmological terms: "Heaven was a quadripartite and multilevel region supported by four divine brothers. . . . The earth was also divided into four parts. . . . The center and each world-direction had its particular god, and each was also associated with a color" (Marcus 1973: 912).[15] She suggests that the rulers of Copan and Seibal

15. A great deal of archaeological, ethnohistoric, and ethnographic research has documented the endurance of a quadripartite cosmology among Maya groups.

FIGURE 17. Seibal Stela 10. (From Graham 1996: 31.)

viewed the Classic Maya geopolitical landscape as a quadripartite order, rooted in four major cities located in different quadrants of the lowlands. In drawing links between these representations of political order and the physical order of geopolitics, Marcus proposes that both stelae provide an approximately mimetic representation of the contemporary geopolitical landscape clothed in cosmological terms—that the Maya lowlands were in fact organized around a limited number of major regional-states centered in large capitals. Thus, the shift in the cast of characters from the Copan stela in A.D. 732 to the Seibal stela in A.D. 849 reflects a waning of Copan and Palenque as geopolitical forces and the rise of Seibal and Motul de San José.

If we are willing to grant the stela interpretive status as mimetic renderings of the "real" geopolitical order, then Marcus's interpretation seems highly plausible. However, as Stephen Houston and others (including Marcus herself) have cogently argued, the glyphic record of the Classic Maya should not be accepted at face value; Maya records are "without question stereotyped, restricted in scope, and edited for appropriate religious and political content" (Houston 1993: 95; see also Marcus 1992). Given the highly stylized geographic associations between directions and places represented in the Copan stela (it is difficult to find a locus from which Copan might lie in the east and Tikal in the west), there is good reason to suggest that these inscriptions were not mimetic renderings of contemporary politics but rather imagined representations of geopolitical landscapes. Such an interpretive stance requires that, instead of reading the inscriptions as snapshots of a contemporary array of dominant polities, they represent the cartographic imaginations of their authors (or sponsors).

What is interesting about the representation of the Maya spatial imagination, as rendered in Stela A from Copan and Stela 10 from Seibal, is not their reiteration of a physical geopolitical order but rather the divergence of the real from the represented. What Marcus takes to be a rather mimetic relationship between physical space and representation I suggest was quite possibly a proscriptive spatial fantasy central to the reproduction of political authority. In this sense, Waxaklahun-Ubah-K'awil's and Wat'ul's representations provide us with a cartography of Classic lowland Maya geopolitics, whose contours were most profoundly shaped by the political ambitions of regimes. Thus, both stelae may be read as instrumental elements in the broader project of producing geopolitical landscapes. They operated, however, not by enforcing specific contours in physical space or by evoking an understanding of authority relations but by advancing a particular imagination of the proper political community. In this respect, the Copan and Seibal stelae are quite similar to the Sumerian King List that opened this chapter.

The Sumerian King List describes how, after a massive flood swept over the earth, kingship was again lowered from heaven to the city of Kish.[16] Thereafter, kingship moved 19 times between 10 different places, nine cities, and one "horde."[17] At the end of each cycle, the reigning Great Power was vanquished in battle, and kingship was carried away to a new locus from which authority over all of Sumer and Akkad was exercised. The most extensive versions of the list end with 14 kings of the city of Isin. For a short time after the fall of the Third Dynasty of Ur (ca. 2017–1897 B.C.), the first six kings of Isin outlined in the King List (the dynasty of Ishbi-Irra) did indeed rule the central provinces of Sumer and Akkad. However, their sovereignty was effectively ended when military defeat at the hands of Larsa inaugurated an extended era of confrontation and contest organized around a multipolar macropolitical order, wherein major cities such as Uruk, Kish, and Babylon repudiated Isin's claims to regional kingship (Hallo and Simpson 1971: 93). In this respect, the Sumerian King List provides a geopolitical genealogy for the rulers of Isin, the hegemonic power in southern Mesopotamia at the most likely time of the texts' composition.

There is much disagreement over how to describe the narrative form of the King List. As Piotr Michalowski (1983) has noted, it is a difficult text to classify as either chronicle or annal, two forms of historical narrative with strong traditions in ancient Mesopotamia. And to describe it as a list over-privileges the unique lines of personal names at the expense of the intensely repetitive structure of the whole. The value of the text as historical record has itself been subject to some criticism because even the less overtly mythic portions of the King List rearrange contemporary rivals into lineal successors. Furthermore, the King List glosses over the fractious rivalries between city-states that characterized the geopolitical reality for the great majority of the third and early second millennia B.C. in order to present Sumer and Akkad as an indivisible political whole. The narrative of the text thus simultaneously flattens and stretches the Mesopotamian geopolitical landscape, compressing rulers known to have been contem-

16. The antediluvian section of the Sumerian King List is widely regarded as a later addition to the texts (Michalowski 1983).

17. The horde of "Gutium" is usually thought of as a tribal group, though there is considerable debate as to its role in the collapse of the Akkadian empire.

poraries into a sequential order and stretching the territory of each "Great Power" to include all of Sumer and Akkad (Michalowski 1983: 243).

The basic premise that organizes the Sumerian King List is that only one ruler in one city should exercise sovereignty over southern Mesopotamia. The macropolitical order is presented as coherent and monolithic because kingship invests suzerainty over all of southern Mesopotamia in a single ruler rooted in a single place. Although the center of this polity migrated, even to places such as Mari and Susa outside of southern Mesopotamia, the essential unity of Sumer and Akkad was preserved. The idealized macropolitical order represented by the Sumerian King List privileged the relatively rare moments in early Mesopotamian history when rulers such as Sargon of Akkade and Ur-Namma of Ur succeeded in assembling authority over all of southern Mesopotamia. Lost are the extended periods of political contest organized around vying city-states and rival coalitions that fractured southern Mesopotamia during most of the third and early second millennia B.C. The only reference to such a multipolar world in the Sumerian King List is found following the reign of Naram-Sin of Agade in a single line, dripping with caustic irony, that reads "Who was king? Who was not king?" (Jacobsen 1939: vii.1).

What is immediately clear from comparison with other historical sources is that the authors of the Sumerian King List consistently rearranged kings and dynasties such that contemporaries and rivals were recast as predecessors and descendants. Given the mechanical nature of the narrative style and the focus on recasting a geopolitical order as an enduring historical one, it seems that the uniqueness of the Sumerian King List follows from the fact that it is not part of a historiographic tradition but rather, like Stela A from Copan and Stela 10 from Seibal, a cartographic one. In order to maintain the regularity of the overall historical pattern that allowed for only one ruler of Sumer and Akkad, spatial relations between contemporaries and rivals were recast as a single temporal line of succession. This historical method should sound familiar because it is the same willingness to hold space absolute in order to recast contemporaries as ancestors that lies at the heart of social evolutionist absolute ontologies of space.

Authority and Geopolitical Landscapes

The foregoing account of the Classic Maya geopolitical landscape suggests that, instead of looking for essential connections between political form and landscape, we need to seek out the practical interconnections that linked sites in ways that shifted over time. Hence Martin and Grube's analysis opens an important window on Maya political landscapes by eschewing static reconstructions in favor of historical recreations of shifting relations of authority and subservience produced by a set of interconnected polities. What is perhaps most interesting about Martin and Grube's account is that they do not presume the relations of domination and subordinance to be evidence of a singular absolute rule of Classic Maya political organization. Instead, they search for the sources of such relationships and practices defined within the political activities of rulers embedded within a specific place and time. This is not to suggest that no patterns exist in Classic Maya spatial data; rather, those patterns that do emerge should not be attributed to rigid sets of spatial forms that inhere in political types. Instead, the geopolitical landscape, in all its dimensions, must be understood as actively produced within the political practices that constitute relations of authority among polities in an ecumene.

Without an account of the constitution of authority through the production of geopolitical landscapes, it is difficult to describe an appropriate framework for interpolity relationships. Although both sides during the bipolar days of the Cold War were content to frame their rivalry in millennial terms as a question of who was on the "right" or the "wrong" side of history, in a multipolar world such historical delusions appear both absurdly reductive (because complex heterogeneous orders must be distilled into antiquated us/them rivalries) and hopelessly naïve (because the "right" side of history is currently not a problem of moral rectitude but of simple endurance). The canonical Cold War representation of the geopolitical order divided the globe into three parts. International relations were represented as a struggle between the First World (Euro-America) and the Second World (the Soviet bloc) that was played out largely in the emerging nations of the Third World. The origins of this tripartite "metageography" lie not in any clear sense of the spaces created by international politics but rather in the by-products of the historical order of global politics that gave rise to the postwar world (Agnew and Corbridge 1995: 19). The collapse of the Second World and the rapid economic development of parts of the Third World have swept away the tripartite model, destabilizing our conceptualization of the spatial organization of the global political order and encouraging the propagation of new geopolitical maps. One map proposes to replace the three-worlds division with a bipolar opposition of a wealthy North with an impoverished South (Lewis and Wigen 1997: x). But such an equatorial division has the great disadvantage of lumping together global rivals, such as the United States,

Japan, China, Europe, and Russia, while opposing traditional allies, such as the United Kingdom and Australia. A less literal post–Cold War geography divides the world into centers, semiperipheries, and peripheries using terminology derived from Wallerstein's (1974: 355) description of the modern world system. Rooted in a historical account of its own emergence since the fifteenth century, world systems analysis has proven a robust model of the expansion of European capitalism. But the failure of Western-style capitalism to take firm root in much of the Second World and the rise of a new global economy of capital and resource flows that have led to the rise of new locales and the decline old ones in both center and periphery have exposed the limited ability of Wallerstein's rather mechanical geography to move beyond the modern (see Sassen 1994; Zukin 1991). As Ulf Hannerz has suggested, "World cultural process, it appears, has a much more intricate organization of diversity than is allowed in a picture of a center/periphery structure with just a handful of all-purpose centers" (1992: 221).

Samuel Huntington (1993, 1996) has forwarded the most apocalyptic vision of the emerging world order, positing that our new geopolitical landscape will coalesce around cultural fault lines rather than ideological or economic ones. Post–Cold War geopolitical conflict, he argues, will hinge on clashes between a handful of age-old "civilizations," differentiated from each other by history, language, culture, tradition, and religion. Huntington's is perhaps the most potentially damaging of all the newly imagined maps of the global political landscape because he misreads the dynamics of culture as coterminous with, and entirely bounded by, the dynamics of geopolitics. This is indefensible from a number of points of view. From a geographical perspective, Huntington's model suffers from what Martin Lewis and Kären Wigen (1997: 11) refer to as a "jigsaw-puzzle view of the world," neatly ordered into sharply bound units without contested frontiers. The result is an account of the world that is fixed and entirely stable not just today but deep into the past as well. From an anthropological standpoint, Huntington's mosaic geography of world cultures appears rather quaint in its revival of a long-outdated model where cultures are easily mapped in place. Such a rendering fits neither the dynamic cultural flows of the modern world (what Hannerz has termed "the global ecumene") nor the geopolitical practices of early complex polities (Hannerz 1992: 217–67; see also Appadurai 1996; Rosaldo 1989). As the case of the Classic Maya suggests, multilateral political relationships cannot be easily mapped along enduring "cleavage planes," to use Marcus's terms. Political divisions and alliances, such as those created by Calakmul, are decidedly opportunistic and predicated on shifting relationships of authority, not on the fundamental encoding of metahistory. Thus, in understanding the emergence and reproduction of authority within inter-polity practices, landscapes cannot be held constant as unchanging elements of evolving structural forms.

Like the fractured, compressed cartography of the Sumerian King List, Huntington's model of contemporary geopolitics is not meant to describe actual fault lines but to actively produce them in specific places. Huntington is intent on reading the postcolonial reduction in direct Western authority and a revitalized cultural self-awareness of the formerly colonized as a recipe for cultural battle. The cannily apocalyptic phrase "Clash of Civilizations" is an attempt to produce a new geopolitics that is somehow deeper than politics, that can be rooted in divisions in humanity that are essential and metageographic and thus encourage nostalgia for old forms of colonialism. Classic Maya polities were engaged in a very similar project. That is, they attempted to produce on the ground, in evocative aesthetics, and in imagined cartographies a complete sense of the enduring authority of contemporary regimes. Central to such a project, conceived on a geopolitical scale, is the polity itself. Described as a coherent and complete unit of political action on the world stage, the polity can itself be understood as a political landscape, as a spatialized set of political practices dedicated to producing and reproducing authority in relationships between subjects and regimes.

CHAPTER 4

Polities

---

*The Scribe Inena communicating to his lord, the Scribe of the Trea-
sury Qa-g[abu]. . . . Another communication to my lord to [wit: We]
have finished letting the Bedouin tribes of Edom pass the Fortress [of]
Mer-[ne]-Ptah Hotep-hir-Maat . . . to the pools of Per-Atum.*

Report of an Egyptian Frontier Official, late Nineteenth
Dynasty (ca. 1295–1188 B.C.; from J. A. Wilson,
"The Report of a Frontier Official")

*The countries of Khor and Kush, The land of Egypt You [god Aten]
set every man in his place.*

Egyptian Hymn to Aten, late Eighteenth Dynasty (ca. 1552–1295
or 1314 B.C.; from M. Lichtheim, *Ancient Egyptian Literature*)

Perhaps the most remarkable single artifact bearing on the formation of
an early complex polity is the shield-shaped slate "Palette of Narmer" that
was discovered in the ruins of a temple at Hierakonpolis in the Nile River
valley of southern Egypt (fig. 18). On one side of the palette, carved in
low relief, a king, wearing the bulb-tipped white crown of Upper Egypt
and identified by the Horus name "Narmer," stands poised with mace in
hand ready to smite a captive (perhaps a rival ruler) delivered by Horus
from the Delta region of Lower Egypt (Kemp 1989: 42; Aldred 1984: 81).
Below this scene lie two fallen figures accompanied by the outline of a
fortified town and a ribbon-shaped emblem.[1] In the upper register on the

---

1. The latter has been described in Aldred 1984: 81 as symbolizing the gazelle traps characteristic of the Sinai Peninsula.

FIGURE 18. The Palette of Narmer. Late Predynastic, ca. 3150–3125 B.C. Recovered at Hierakonpolis. Slate, H:63.5cm. Egyptian Museum, Cairo. (Photo courtesy of Scala/Art Resource). **a.** Back. **b.** Front.

other side of the palette, Narmer, now wearing the red crown of Lower Egypt and accompanied by four distinctive standards, surveys two rows of bound and decapitated enemies. At the bottom of the image, Narmer, rendered in the form of a bull, razes a fortified town, treading his enemy underfoot. In between these scenes of destruction, serpentine heads of lionesses intertwine to symbolize the unification of Upper and Lower Egypt into a single polity.

What is remarkable about the Palette of Narmer is the way in which it simultaneously maps the territorial claims of a unified Pharaonic polity (encompassing the territories of Upper and Lower Egypt), represents the polity in the body of the new king, and attempts to generate an imagined community through syncretic symbols of political coalescence (a syncretism that would further play out during the First Dynasty in such motifs as the double crown). In its remarkably succinct way, the Palette of Narmer addressed the central spatial problems for constituting authority within polities—the delineation of a bounded territory within which a sovereign regime rules a community of subjects integrated by a shared sense of identity that binds them together in place. The formation of the Pharaonic polity is framed, at a very early moment in dynastic history, as coterminous with the production of a distinct political landscape.[2]

Following Anthony D. Smith's definition of the "nation" as "a named community of history and culture, possessing a unified territory, economy, mass education system and common legal rights" (1996: 107) it could fairly be argued that the Palette of Narmer represents an initial statement in a project of dynastic Egyptian nation building. The palette itself describes the unification of two named communities into a territorial polity; within a few centuries, Old Kingdom Pharaohs would sit at the epicenter of a large administrative apparatus concerned with the regulation of religion, social order, and economy.[3] I do not raise the term "nation"[4] as a salvo in the essentially typological dispute over whether such polities existed in antiquity or are unique to the modern world.[5] Nor do I intend

2. A. J. Spencer (1993: 53) is no doubt correct in arguing that the Palette of Narmer should be understood as one part of a gradual process of political accretion in the Nile valley rather than a snapshot of a single moment of unification; however, this observation does not decrease the complexity of the palette's representational strategy.

3. John Baines (1995: 3) argues quite forcefully that ancient Egypt can be described as "the first large 'nation state.'"

4. Charles Tilly (1975: 6) has described the concept of "nation" as "one of the most puzzling and tendentious items in the political lexicon."

5. It is worth noting that writers who have argued against extending the nation beyond the historical confines of the modern era make their arguments on extremely shaky archae-

to argue from the case of Egypt that early complex polities were invariably predicated on the close mapping of territory and cultural identity. Rather, this chapter examines how early complex polities were produced as delimited political communities within practical relationships between regimes and subjects. These relationships, I argue, were constituted within the experience, perception, and imagination of political landscapes.

Current debates over nations and nationalism are relevant here, however, because they provide an outline of the varied approaches to theorizing the relationship between subjects, regimes, and polities—between the forms that order territories and the affective ties that bind political communities in place. Nations are generally described as comprising both a historically and a geographically unique field of political practice. Historically, they have typically been defined as formations unique to modern political practice, severed from the pre-modern by an awakened consciousness of a one-to-one correspondence between collective identification and the specificity of governmental institutions. Geographically, nations are discrete territories incorporated under a sovereign regime—places predicated on their differentiation within the geopolitical sphere from neighboring polities by thresholds of varying specificity policed with varying intensity. Nations are also presumed to encompass "imagined communities," in Benedict Anderson's (1983) useful phrase, where commonalities of sentiment and history create attachments to place among a community of subjects.

Polity, Nation, and Landscape

The primary interest of nationalist accounts of the nation is to provide a deep and defensible link between people and territory. As such they tend to be dominated by subjectivist spatial ontologies where national terri-

---

ological foundations. For example, Anthony Giddens's (1985: 50–51) assertion that there was no conception of bounded territory in antiquity—marked locales that divide internal space of the polity from the external world—is simply unfounded, as the epigraphs that open this chapter suggest. Although it is worth noting the error, particularly because it has been repeated so often in various contexts as to appear canonical (Anderson 1983: 7; Gellner 1994: 35–36; Hobsbawm 1990; Kohn 1962), very little actually hangs in the balance in correcting it; the only work such pretensions to exclusivity appear to do for modern political theorists is to sustain the illusion that the historical starting point of their genealogies (usually sometime after A.D. 1700) is less than arbitrary. This is at least a representational problem and at most a methodological one and so bears correction but not extensive treatment.

tories are understood to be direct geographic expressions of enduring historical ties among a collectivity rooted in blood (race), brain (national spirit, language), or belief (religion).[6] Although subjectivist accounts of nation were strongly criticized in the late nineteenth and early twentieth centuries for promoting essentialized accounts of identity and community formation, they have enjoyed a modest revival in the political theories of Steven Grosby and Anthony D. Smith. These neo-subjectivist accounts of the nation rely on highly problematic understandings of relations between subjects, political regimes, and place in early complex polities and so warrant some discussion here.

We can see the neo-subjectivist account of the space of the nation most clearly in Grosby's (1995) account of territoriality. Territory, Grosby argues, is a primordial element of human society that arises from the life-sustaining needs of both the individual (as a discrete set of available physical nutrients) and the social collective (as a locus "for those memories and psychic patterns necessary for the ordering of life" [ibid., 158]). I am sympathetic to Grosby's attack on modernist claims to the historical uniqueness of the nation as a form of political organization (see his excellent 1997 study of boundaries in ancient Armenia, Edom, and Ara). But his effort to root an understanding of territory in its capacity to provide physical and psychic sustenance for the collectivity is ultimately unconvincing both as political theory and as a reading of ancient political history. What Grosby is trying to do is to reestablish a direct relation between people and place as read through sociobiological renderings of organismic requirements. But this theorization overlooks the essential differences between catchment areas—the geographic area that provides subsistence resources to a given population—and political territories. These do not map easily onto each other, either in the modern or in the ancient worlds. The report of the Egyptian frontier official that opens this chapter provides a good case in point. The pastoral herders allowed to pass into Egyptian territory are clearly not members of the Pharaonic polity, yet such flocks did contribute to its subsistence economy, to the "physical nutrients" of the Egyptian political community.[7] Catchment area and polity cannot be regarded as coterminous spaces.

As political theory, Grosby's approach to the territory of polities al-

6. For an overview of nineteenth-century currents, see Renan 1996.

7. Studies of pastoral nomads in ancient southwest Asia consistently point to their exteriority to the polity despite the centrality of their produce to complementing the agricultural economies of urban centers (see, e.g., Cribb 1991; Khoury and Kostiner 1990).

lows for only a relatively limited set of explanations of historical transformations in territorial form and extent. Ecological change or alterations
in the nutrient requirements of populations are the only clear determinants that might explain changes in attachments between people and place.
What this bio-substantivist framing of the problem can never answer is
why political communities took the form that they did, where they did,
by what means, and to what ends. In other words, noting that people
have to get their food from a specific place and that often this subsistence
attachment can be culturally elaborated simply does not suffice as an account of how polities are produced through spatial practices that instan-
tiate a sense of territorial belonging and political sovereignty. As Robert
Sack (1986: 30) has argued, territoriality is a strategy, a device for defining
and maintaining spatially delimited organizations; it is thus "a product
of social context."

In a more phenomenological variant of the neo-subjectivist reading of
nation, Smith has suggested that the cohesion of nations within communities of sentiment is created out of "lines of cultural affinity embodied in myths, memories, symbols and values" (1991: 29). Although Smith
is undoubtedly correct in placing the debate over national formation in
the domain of cultural production rather than subsistence economy, his
rather ungrounded account of the polity tends to obscure any sense of
place, leaving us rather unclear as to how values, myths and memories became rooted in particular locations and forms. For example, the literary
critic Nona Balakian once remarked, in relation to the Armenian diaspora's
imagining of political community, "We [Armenians] have a dream instead
of a country. . . . The more our geography shrinks, the more our imaginations expand" (quoted in Balakian 1997: 138). Yet this imagination is
always clearly articulated with a particular place in eastern Anatolia that
provides a materiality to the Armenian diaspora's national sense of loss.
That is, the imagined community is only intelligible when that imagination is embedded in place. What is missing from the neo-subjectivist spatial ontology of both Grosby and Smith is a sense of the polity as a horizon of action, a field of political practices that produce boundaries,
frontiers, and places steeped in memories within a landscape that aspires
to cohere as a locale of sovereign authority. In other words, how do polities emerge as landscapes—as experientially discrete territories, as places
perceived to evoke enduring commitments of people to land, and as imagined accounts of the sources of enduring attachments between subjects
and regimes rooted in space and time? To attempt to understand processes
of, for example, Armenian national formation separate from an under-

standing of the Armenian highlands as a political landscape is to displace the very stake of nationalist practices—the homeland. This is no less the case for the ancient world than it is for the modern.

The report of the Egyptian frontier official and the hymn to Aten that opened this chapter provide good examples of the spatiality of the practices of polity formation. In the communiqué from the frontier, we can see the political landscape constituting the polity in a material sense, where the Pharaonic regime during the New Kingdom period was predicated on the surveillance of formalized borders by a series of fortified posts that regulated transit into and through Egyptian territory. In a more imagined sense, the hymn to Aten suggests that this regulation of political boundaries was in part predicated on a sense of politico-cultural difference such that distinct social groups were rooted at creation in distinct places. Thus the integrity of the Egyptian polity in the late second millennium B.C. was framed as a product of theogony maintained through political action.

Although investigations of early complex societies have typically viewed subject and regime as binary components of polities linked by either coercion or consent, I do not define them as mutually exclusive social positions. Because the constitution of regimes is the primary focus of the next chapter, I will touch on the concept here only to offer a working definition suitable for the present analysis. Regimes are located in the intersection of a power elite that controls critical institutions of governance (the political apparatus) and grassroots coalitions of "like-minded" subjects committed to sociopolitical reproduction. Regimes should thus not be understood as the dialectical opponent of the subject, because subjects are active elements in the reproduction of their authority. Political subjects are far more difficult to define.[8] For the sake of simplicity, I will use the term to refer to individuals and groups who must respond to a regime's demands and who recognize, or refuse, the legitimacy of these dictates. The term thus embraces not only those who are potential parts of the regime's coalition but also all those who fall under a regime's effective sovereignty. Sovereignty refers to the establishment of a governmental apparatus as the final authority within a polity and therefore entails both the definition of a territorial extent beyond which commands go unenforced and unheeded and the integration of discrete locales into a singular political community (Hinsley 1986: 26; Hoffman 1998).

8. See Žižek 1999 for one of the more recent considerations of the problematic nature of the political subject, but compare Foucault 1978, 1982.

In order to further develop an account of the experiential, perceptual, and imagined landscapes of polities, this chapter focuses on the kingdom of Urartu. Urartian regimes ruled eastern Anatolia and much of southern Caucasia during the early first millennium B.C. (ca. 850–643 B.C.). I will first examine Urartian efforts to redefine the sources of the political landscape in the imagination of subjects through royal declarative inscriptions. In addition to advancing a new representational aesthetic, Urartian authorities also attempted to define a new perceptual sense of the built environment in conquered regions, one that simultaneously razed preexisting attachments to the land and capitalized on traditional perceptions of place. These transformations in spatial perception and imagination were accompanied by a dramatic alteration of the physical relationships between subjects and the built environments of the political apparatus, reconstructing the experience of landscape to promote the political goals of the ruling regime.

Landscape and Polity in Urartu

The kingdom of Biainili, known to its contemporaries the Assyrians (and hence modern scholarship) as Urartu, appears to have emerged in eastern Anatolia from a group of local polities during the late second and early first millennia B.C.[9] Between the mid-ninth and late eighth centuries B.C., Urartu embarked on a program of imperial expansion, conquering rivals from the headwaters of the Euphrates to the south shore of Lake Urmia and from the foothills of the Taurus mountains to the intermontane plains of southern Caucasia (fig. 19). It is this northern province of the Urartian kingdom, an area centered on the Ararat plain in what is today the Republic of Armenia, that provides the primary site for much of the following discussion. Although a presence north of the Araxes river since the reign of King Ishpuini in the late ninth century B.C, the Urartian occupation of southern Caucasia did not begin in earnest until the second decade of the eighth century B.C. when King Argishti I formalized the

9. During the late second millennium B.C., Assyrian inscriptions variously refer to this northern group of polities as "Ur(u)atri" (beginning in the reign of Shalmaneser I) or "Nairi" (beginning with the reign of Tukulti-Ninurta I). The exact referents of "Ur(u)atri" and "Nairi" are not entirely understood. In some inscriptions (e.g., Grayson 1987: A.0.77.1) they appear to be general geographic designations, whereas in others (e.g., Grayson 1987: A.0.87.1) they appear to denote polities or confederacies united in resistance to Assyria (see Salvini 1967).

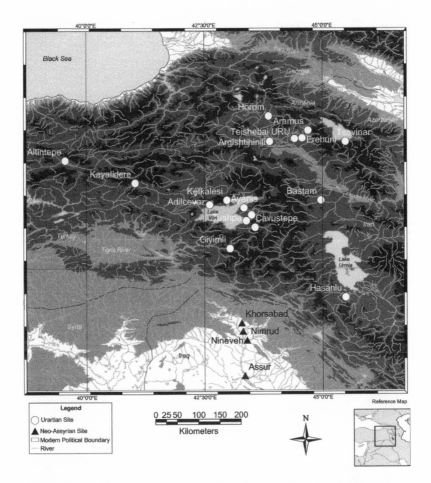

FIGURE 19. A map of eastern Anatolia and southern Caucasia showing major Urartian sites of the ninth to seventh centuries B.C. (Map source: ESRI Data & Map CD.)

kingdom's military conquests through an extensive program of construction focused in the Ararat plain (fig. 20).[10]

The era of high Urartian imperial expansion was brought to a close by a series of military defeats in the late eighth century B.C. Urartian military and diplomatic incursions into the southern Urmia basin of northwestern Iran provoked King Sargon II to reassert an Assyrian presence

10. An inscription of King Ishpuini was recently found in the Zangezur region of southern Armenia (Hmayakyan, Igumnov, and Karagyozyan 1996).

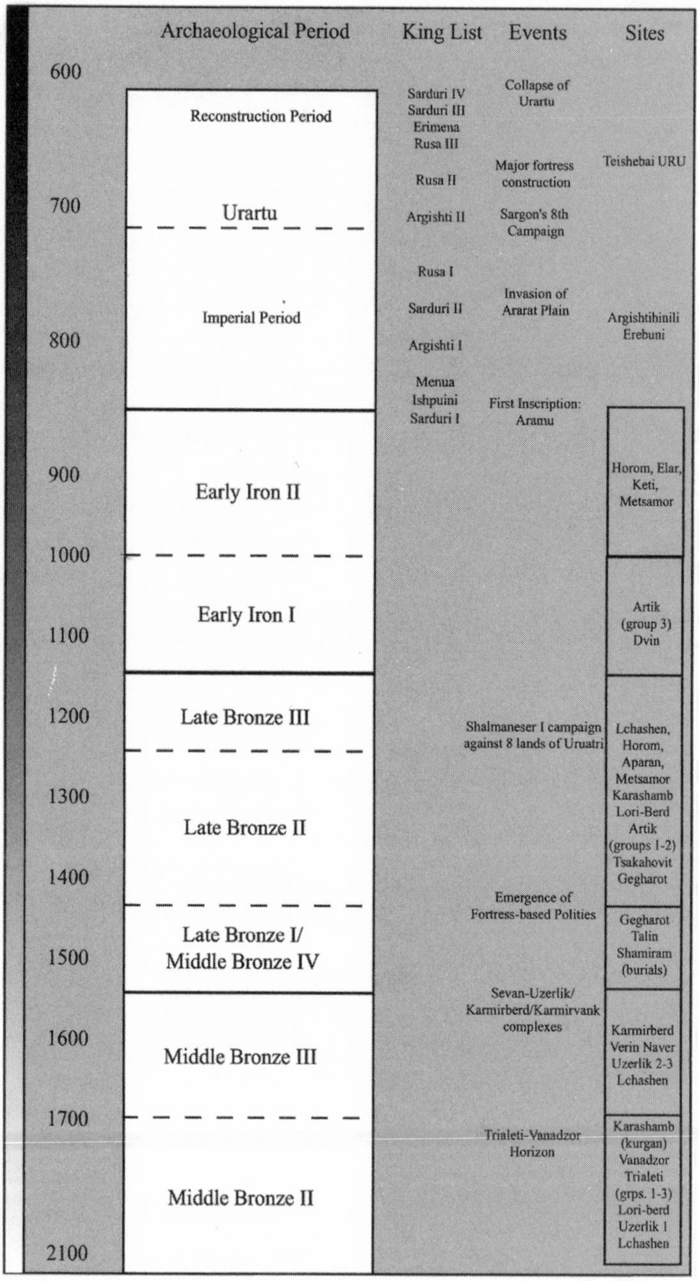

Date (BC)	Archaeological Period	King List	Events	Sites
600	Reconstruction Period	Sarduri IV / Sarduri III / Erimena / Rusa III	Collapse of Urartu	
	Urartu	Rusa II	Major fortress construction	Teishebai URU
700		Argishti II	Sargon's 8th Campaign	
		Rusa I		
	Imperial Period	Sarduri II	Invasion of Ararat Plain	Argishtihinili / Erebuni
800		Argishti I		
		Menua / Ishpuini / Sarduri I	First Inscription: Aramu	
900	Early Iron II			Horom, Elar, Keti, Metsamor
1000				
	Early Iron I			Artik (group 3) / Dvin
1100				
1200	Late Bronze III		Shalmaneser I campaign against 8 lands of Uruatri	Lchashen, Horom, Aparan, Metsamor Karashamb Lori-Berd Artik (groups 1-2) Tsakahovit Gegharot
1300	Late Bronze II			
1400			Emergence of Fortress-based Polities	
	Late Bronze I/ Middle Bronze IV			Gegharot Talin Shamiram (burials)
1500			Sevan-Uzerlik/ Karmirberd/Karmirvank complexes	
1600	Middle Bronze III			Karmirberd Verin Naver Uzerlik 2-3 Lchashen
1700			Trialeti-Vanadzor Horizon	Karashamb (kurgan) Vanadzor Trialeti (grps. 1-3) Lori-berd Uzerlik 1 Lchashen
	Middle Bronze II			
2100				

FIGURE 20. A periodization of Bronze and Iron Age southern Caucasia and a chronology of the Urartian kings. (After Avetisyan, Badalyan, and Smith 2000; Salvini 1995.)

in the region. His campaign climaxed in the defeat of an Urartian army led by King Rusa I.[11] Assyrian intelligence reports indicate that Urartu was also attacked at this time by Cimmerians crossing into Caucasia from the Eurasian steppe and further destabilized by an insurrection within the Urartian ruling elite that threatened the royal dynasty (Lanfranchi and Parpola 1990: nos. 91, 92). Rusa I succeeded in deflecting the Cimmerians and quelling the rebellion, thus preserving the dynasty, but Urartu's era of expansion came to an end, its imperial designs checked by Assyria in the south and by new populations moving into Caucasia from the north.

The historical record for the succeeding era of Urartian political reconstruction is not as rich as that of the preceding imperial era. But though the historical sources are more reticent, the archaeological record is substantial, indicating a reconsolidation of much of Urartu's territory, a resurgence of Urartian resolve to challenge Assyrian pretensions in the highlands, and a reinvigoration of the authority of the Urartian regime. The reign of Rusa II represents the apogee of this reconstruction. Thanks to foundation inscriptions, we know that five major fortresses, accomplished on a massive scale, are directly attributable to his reign, including Karmir-Blur on the Ararat plain within the western precincts of modern Yerevan (Oganesian 1955; Piotrovskii 1955).

Dynastic succession following Rusa II is unclear, leaving some confusion over the last rulers of the empire and the dating of collapse. The fate of Urartu and its possessions in southern Caucasia during the late seventh century B.C. is not well understood. Boris Piotrovskii has dated the final collapse of Urartu to 590 or 585 B.C. based largely on a biblical reference, but this chronology is generally thought to be overextended.[12] An inscription of Ashurbanipal, dated to 643 B.C., records the submission of the Urartian king "Ishtar-duri" to the Assyrians (Sarduri III or IV). Although not an entirely satisfactory date for collapse, afterward Urartu was never again a significant force in the geopolitics of southwest Asia.

Two general models have been developed to characterize the political organization of the Urartian kingdom. The first suggests that Urartu was a highly centralized polity led by a king of singular authority who actively redistributed resources and labor from the peripheries to the regional centers (Adontz 1946; Melikishvili 1951). A more recent model posits a less

11. Sargon's account of his eighth campaign has proven a rich source for geographical analyses of the Urmia basin (e.g., Levine 1977; Zimansky 1990).

12. On the long chronology, see Piotrovskii 1959, 1969. The biblical reference that Piotrovskii cites is Jeremiah 51:27. For critiques, see Kroll 1984 and Zimansky 1995.

centralized system of authority. In this model, the Urartian king exercised control over a mosaic of provinces ruled by local administrators who were required to supply troops for the king's campaigns. Through the king's direct authority over the political bureaucracy, the potential for rebellion against the center was minimized (Zimansky 1985: 94). The first model emphasizes the regularized use of coercive instruments of centralized state power to establish relations of radical inequality between subjects and the Urartian regime; the second model attends to the institutional organization of the Urartian state, positing a much more decentralized pattern of governance shaped by the rugged regional topography. These models are not necessarily mutually exclusive, and each can muster support from the archaeological and epigraphic records. However, neither articulate an account of the production of Urartu as a polity, a political landscape defined through the practical relationships of regime and subjects.

Before moving into a consideration of the Urartian polity and the landscape it forged, it is important to note two structural dimensions of the following discussion. First, although the study begins with a geographically broad consideration of Urartian epigraphic representations of the built environment, the subsequent studies of perception and experience focus more narrowly on the Urartian occupation of the Ararat plain. By attending to a specific province rather than the whole of the polity, we are able to more closely define the practical relationships between regime and subjects that arose within the experience and perception of a politically produced landscape. Second, the reader will note that this chapter provides a more direct and intensive engagement with the archaeological and epigraphic data than the previous chapter on the Classic Maya. The reason for this is simply my more immediate involvement with the study of Urartu and its predecessors through my ongoing fieldwork in southern Caucasia. I hope that the reader will forgive this slight shift in voice.

IMAGINATION

In numerous display inscriptions carved into exposed rock faces or stone stelae, the kings of Urartu called attention to the personal heroism of their building activities: "The earth was wilderness; nothing was built there; out of the river I built four canals, vineyards, and I planted the orchards, I accomplished many heroic deeds there. Argishti, son of Menua, powerful king, great king, King of the lands of Biainili, ruler of the city of Tushpa" (Melikishvili 1960: #137). Urartian monarchs described construction activities as episodes of conquest, advancing a claim to political legitimacy based on the power of the king to subdue the "wilderness" and call forth an ordered landscape. This ideological program staked the political order on what might be called the tectonic charisma of individual kings, by which I mean their unique capacity (derived from supernatural or other sources) to transform an "uncivilized" or undisciplined natural world and make it fit for human social life through construction (see Kus 1989: 142).[13] The built environment is represented as historically specific in its production and explicitly political in its sources. These display inscriptions also provide an account of the simultaneous destruction of the preexisting political landscape and production of a new Urartian landscape within the imaginative encounter between a political regime and political subjects.

The carving of display inscriptions appears to have been a royal prerogative in Urartu. As Paul Zimansky notes, "Display inscriptions were intended for the glorification of the ruling monarch, and consequently the name of the king and his patronymic were essential elements" (1985: 50). It is largely thanks to the parentage statements contained in these texts that we can trace the dynastic succession of Urartian kings from the mid-ninth century B.C. through the reign of Rusa II two centuries later.[14] The prevalence of patronymic details in all but the most damaged display inscriptions suggests that we can safely conclude that they were political creations closely linked to the institution of kingship.

There are two general types of Urartian inscribed representations of the built environment. In the first group are simple founding inscriptions associated with the construction of fortresses and individual buildings, such as houses, granaries, and temples. In general they are brief texts, giving the name of the king responsible for the construction and his pedigree (son of . . .). For example: "Ishpuini, son of Sarduri, built this house" (Melikishvili 1960: nos. 4–10, 13; Smith 2000; see also Melikishvili 1960: nos. 25, 26, 29, 43–56, 59–62, 88–91, 162–66; Melikishvili 1971: nos. 374–80, 391–95, 420–23). The earliest known inscription of an Urartian king was simply a foundation inscription preceded by an elaborate titulary:[15] "Inscription of Sarduri, son of Lutibri, great king, mighty king, king of

13. By charisma I mean "a quality of an individual personality by virtue of which he is set apart from ordinary men and treated as endowed with supernatural, superhuman, or at least specifically exceptional powers or qualities" (Weber 1968: 48).

14. Inadequate documentation from the reigns of the kings who followed Rusa II limit our ability to reconstruct the final years of Urartian power.

15. Although Melikishvili 1960: no. 1 describes the inscription as written in "the Assyrian dialect of Akkadian," Piotr Michalowski (personal communication 1998) suggests the text was written in Akkadian.

the universe, king of Nairi, king who has no equal, wonderful pastor, fearless in battle, king who subdues the intractable. (I), Sarduri, son of Lutibri, king of kings, who obtained tribute from all kings. So speaks Sarduri, son of Lutibri, I brought this stone from the city of Alniunu. I erected this wall."[16] Although this text may seem strangely anticlimactic to the modern ear (this wall is impressive, but its construction seems somewhat inconsequential for the "king of the universe"), foundation and construction are rendered as actions that demonstrate the majesty of the king referred to in the titulary. Thus the latter portion of the inscription provides an example supporting the boasts of the former.

The dominant figurative device at work in these very simple declarations is a reduction of the Urartian political regime to the person of the king; all construction is accomplished by and through the king. Although an instrumental role is often given to the deities (particularly Khaldi), it is only through the king that building takes place. As a literal discourse, these texts can be read as event markers. But as figurative renderings of the built environment, the poetics of the texts establish the more profound thesis that the ability to accomplish construction projects, from granaries and houses to complete fortresses, lies solely with the king. This transformational power is the foundation of tectonic charisma.

A second type of epigraphic representation of the built environment is the landscape inscription. These texts, though also recording founding events, are marked by a concern not only to establish the ability of the king to build but also to evoke an impression of his more encompassing power to transform undifferentiated spaces into politically constituted places, as in an inscription of the seventh-century B.C. by King Rusa II:

> To Khaldi, his lord, Rusa, son of Argishti, erected this stela. By the might of Khaldi, Rusa, son of Argishti, speaks: the earth of the plain [or valley] of Kublini was unoccupied (?), nothing was there. As (?) god Khaldi ordered me, I planted this vineyard; I planted here new fields and orchards, I built here a city. I diverted a canal from the river Ildarunia—(its) name "Umeshini". [. . .] Rusa, son of Argishti, speaks: whoever obliterates this inscription, whoever moves (it), whoever removes (it) from (its) place, whoever buries (it) in the earth, whoever throws it in the water, whoever says to another: "I accomplished (all this)," whoever obliterates (my) name (from here) (and) supplies their own, whether he is an Urartian or a barbarian [enemy], let the gods Khaldi, Teisheba, Shivini, (all) of the gods allow neither him nor (his) name, nor (his) family, nor (his) progeny to remain on the earth. (Melikishvili 1960: no. 281)

---

16. This translation is a composite of renderings in Melikishvili 1960: no. 1 and Wilhelm 1986: 101 with additional assistance from Piotr Michalowski (personal communication 1998), for which I extend my thanks. See discussion in Salvini 1995: 34–38.

A similar, though more economical, landscape inscription from the early eighth century B.C. marked the founding of the first Urartian fortress on the Ararat plain:

> By the majesty of god Khaldi, Argishti, son of Menua, built this fortress perfectly; and [gave to it] the name Irpuni (Erebuni); (It was built) for the greatness of Biainili (and) for the humiliation of the enemy lands. Argishti says: The earth was wilderness; I accomplished great deeds there. (Melikishvili 1960: no. 138)

We can identify three dominant figurative processes in the extant landscape inscriptions.[17] Although not every inscription will necessarily include all three elements, we can identify them as aspects of the representational discourse that were mustered in varying combinations or in toto.

Evacuation.   One of the most striking elements of these inscriptions is their relentless emphasis on the emptiness of the landscape prior to the arrival of the Urartian kings. This emptiness can be interpreted either narrowly or broadly. The narrow interpretation suggests that these texts simply refer to the lack of a preceding structure on the immediate site on which Urartian fortresses were constructed. Archaeological excavations tend to bear out suggestions of an Urartian preference for constructing on bedrock rather than atop cultural levels deposited by preceding occupants (see chapter 6). The primary thrust of the narrow interpretation is to understand the Urartian emphasis on emptiness as purely descriptive of the condition of the immediate building site.

A broad interpretation, in contrast, would argue that when the Urartians referred to an empty wilderness prior to their arrival it referred to a more encompassing sense of landscape. Thus, the emptiness emphasized in these inscriptions would be more poetic than descriptive—a figurative description of a locale unincorporated into the more "civilized" world of the Urartian empire. We know from archaeological investigations that numerous polities occupied the highlands of southern Caucasia and eastern Anatolia centuries before Urartian expansion (for example, Badalyan, Smith, and Avetisyan 2003). Furthermore, the regions portrayed in landscape inscriptions as vacant were simultaneously described in military annals as crowded with vanquished foes.[18] Thus, in the broad interpre-

17. Other landscape inscriptions include Melikishvili 1960: nos. 16, 17, 65, 137, 165, 167, 169, 170, 172, 265, 266, 276, 278, 280; Melikishvili 1971: nos. 14, 16, 372, 388, 389, 396, 397, 418, 448, 453, 455; Van Loon 1975.

18. Compare, for example, Argishti I's founding inscription for Erebuni (Melikishvili 1960: no. 138) with a military inscription recovered at the site of Elar, just a few kilometers

tation, Urartian claims to vacancy must be interpreted as primarily figurative rather than purely descriptive statements.

A figurative approach provides a more compelling interpretation of the extant materials, particularly when we note inscriptions that declare entire regions vacant, not just a specific locale. In the inscription of Rusa II above, the entire "plain of Kublini" is described as empty, not just a particular building site. It is important to point out that, even if the texts hold descriptive significance, this in no way mitigates their figurative operation as a prominent recurring tropic element of landscape inscriptions. The fact that the Urartian kings thought it vitally important to emphasize the emptiness of the preexisting landscape, whether descriptive or not, is revealing of the relationship they sought to forge between the political landscape of the imperial apparatus and local understandings of place.

In contrast to political traditions in Mesopotamia, where regimes went to great lengths—both genealogically and geographically—to demonstrate their articulation with a continuous historical legacy, Urartian kings described themselves as filling a political void.[19] By positing an emptiness that preceded Urartu, these inscriptions constructed a powerful, if suspect, opposition between wilderness and empire mediated by the tectonic charisma of the king. Without presuming too far on contemporary cultural constructions of wilderness, the metaphor expressed in this opposition can be drawn out to read the expansion of the Urartian empire as a triumph of the king's power to impose order on the untamed. By defining a preexisting locale as a tabula rasa, rival understandings of place that might compete with, undermine, or question those advanced by the royal regime were excluded.

Reduction.   Agency in construction is firmly located in the king as symbol of the polity. The king is specifically situated within a dynasty through parentage statements that emphasize the historicity of his building activities. Urartian rulers were so concerned that the historical specificity of the act of construction not be obscured that they placed a curse on any descendants who would undo construction or the record of construction.[20]

---

to the north (Ibid., no. 131). Whereas the former describes the region as empty, the latter records the defeat of the land of *Etiuni* and the city of *Darani* of the land of *Uluani*.

19. Examples of Mesopotamian concern with continuity rather than evacuation include Assyrian genealogies that extend into mythic animals and Neo-Babylonian emphases on regular continuity.

20. Curse formulae are well known throughout ancient southwest Asia, but observation of their popularity is not in itself sufficient to forging an understanding of their role in particular contexts.

In other words, it was of paramount importance that a built structure be attributed to a specific king who led Urartu at a specific time rather than simply be acknowledged as a product of a generalized political apparatus.

Integration.   Urartian kings took great care to detail what they built. This concern extended beyond large-scale monuments to include the layout of vineyards and fields. In so doing, the parts were established in preparation for articulation within a larger domain of the empire. The purpose of building, these texts suggest, is not to glorify the king, though it is his charisma that makes the transformation of the landscape possible. Instead, construction is meant to increase the grandeur of the polity—the "greatness of Biainili." In this way, the built environment of a particular locale is integrated as a portion of the kingdom, situated within the larger political whole as a specific place that testifies to the glory of the empire. Fortresses and canals, granaries and vineyards, are given meaning through their inscription in the larger body politic. Particular places are described as meaningful in reference to their integration into the imperial whole.

In Urartian landscape inscriptions, the movement from the evacuation of preexisting communities to the integration of politically constituted places into the empire defines an implicit narrative of triumphal conquest. The king, through his personal heroism, subdues the wilderness, establishing a built environment that is "civilized" by virtue of its inclusion within the constituted political whole. Construction is rendered in emotional tones that describe the transformation of unregulated space into political place as a personal triumph of the king. The potency of the Urartian imagined landscape lay not solely in the compelling quality of its dominant tropes but also in the large-scale transformation the Urartian kings accomplished in the physical landscape, transformations that reordered the spatial perception and experience of the polity. In order to ground our discussion in the spatial particulars of a single case, the remainder of this chapter focuses on the Urartian province centered in the Ararat plain of southern Caucasia.

PERCEPTION

In producing the political landscape of Urartu, the regime of the kings of Biainili attended not only to reimagining the polity through royal inscriptions, but also to reformulating the affective ties between subject and place. The promulgation of an Urartian polity in the Ararat plain was predicated on both the active establishment of new practical relations between subjects and regime and the destruction of preexisting political communities.

Prior to the arrival of Urartian forces in southern Caucasia, the region appears to have been organized into numerous small political communities. Recent research in the Tsakahovit plain of western Armenia suggests that, by the Late Bronze Age of the mid-second millennium B.C., complex polities had emerged in the region, centered in fortified settlements perched atop defensible rock outcrops (see Avetisyan, Badalyan, and Smith 2000; Badalyan, Smith, and Avetisyan 2003; Smith, Badaljan, and Avetissian 1999). These fortresses were only the most conspicuous element of a broad transformation of the regional landscape that projected far into the hinterlands through new irrigation facilities and mortuary architecture. The vast Late Bronze Age cemeteries of the Tsakahovit plain region may well have marked the territorial boundaries of political sovereignty in the area (Badalyan, Smith, and Avetisyan 2003). But, perhaps more immediately, the landscape forged between fortress and cemetery appears to have produced an enduring sense of place in which the apparatus of the living regime and burials of dead subjects provided critical points of reference for inscribing and regularizing the territorial polity. Set against the backdrop of Mt. Aragats, the combined effect of the confluence of the plain and the massif, the rolling cemeteries and the towering fortresses, remains a powerful vista (fig. 21). The Late Bronze Age fortresses on the slope of Mt. Aragats rose up, in effect, from a vast necropolis. Surrounded by the dead and set at the mountain's edge, the fortresses spatially evoke a sense of mediation between the living and the dead, the immediate and the cosmic, providing a sensuous account of political authority that was strongly rooted in place yet most profoundly about transcendence. Thus, current evidence suggests that, as early as the mid-second millennium B.C., the polities of southern Caucasia were assembled in place within an enduring relationship between regime and subjects that tied both to specific sets of local sites. This was the political landscape that Urartu encountered in its campaigns in southern Caucasia during the early eighth century B.C. and attempted to disassemble as it sought to regularize its governance of the region.

A range of Urartian political practices were dedicated to destroying the commitments of now-conquered subjects to preexisting polities. One set of practices centered on obliterating the built environments of prior political communities by demolishing Late Bronze and Early Iron Age fortresses. The Urartian tendency to raze preexisting fortresses is well attested in southern Caucasia, at sites such as Metsamor and Horom North (Badaljan et al. 1992, 1993, 1997; Khanzadian, Mkrtchian, and Parsamian

FIGURE 21. The Late Bronze Age fortress of Tsakahovit (at arrow) against the backdrop of Mt. Aragats. (Photo by author.)

1973). In the case of the latter, a preceding Early Iron Age settlement was partially destroyed and the ground scraped so that the succeeding Urartian walls were built atop levels deposited in the third millennium B.C. This practice of scraping a building site is well attested throughout Urartu, where builders typically removed all preceding occupation levels before building directly on bedrock foundations.

In one sense, this practice of site clearing undoubtedly reflects an approach to construction whereby well-prepared surfaces provided the foundation of choice for Urartian architects and engineers. But the Urartian penchant for dismantling Early Iron Age fortified political centers also reflects a desire to empty conquered regions of the physical vestiges of prior polities—the very same desire that was given voice in the textual records of Urartian conquests of wildernesses and deserted places. The lingering traces of pre-Urartian fortresses provided a rival architectural aesthetic that not only could cast doubt on the Urartian imagined landscape as a political project but could also provide a rival sense of place that might foster an alternative understanding of the polity rooted in a now-vanquished past. Such an alternative vision of the political landscape was made unthinkable through the obliteration of certain evocative places. In other words, the extirpation of preexisting political centers was not only a technique of construction but also a technology of political memory and forgetting.

This approach to breaking down the commitments of subjects to preexisting polities was reinforced by a second, much more brutal set of Urartian political practices. In addition to removing places of political memory from local landscapes, Urartian regimes ripped people out of place, severing the ties between subjects and embedded political traditions through forced deportation from one area of the polity to another.[21] Urartian inscriptions describe systematic programs of population resettlement that uprooted thousands of subjects from their homelands, resettling them in distant areas of the empire. In response to an apparent uprising in southern Caucasia that followed his initial conquest of the area (ca. 785 B.C.), King Argishti I claims to have pacified the region, capturing 19,255 boys, 10,140 warriors, and 23,280 women: "[S]ome of them were killed, others were brought away alive" (Melikishvili 1960: no. 127). Groups from other parts of the empire were subsequently relocated into southern Caucasia, where they formed the core subject population of the new province. As a result, many of Urartu's subjects in the Ararat plain would have had no personal or historical connection to the places and monuments of previous local polities, thereby removing a critical point of resistance to Urartian attempts to establish a reconfigured polity.

The best example of large-scale resettlement comes from the fortress of Erebuni, built by Argishti I as the royal Urartian center in the Ararat plain. In his annals, Argishti I described the defeat of the enemy (Etiuni), the collection of prisoners, the paradoxical emptiness of the region, the construction of the new fortress of Erebuni, and the resettling of prisoners from the Upper Euphrates region:

> The god Khaldi appeared (on the campaign) with his weapons (?). He conquered the land of Etiuni. [. . .] Argishti speaks: I destroyed the land of the city Kikhuni, located on the bank of the lake. I came up to the city of Alishtu; I stole away men and women (from there). For the greatness of Khaldi, Argishti, son of Menua, speaks: I built the city of Irpuni [Erebuni] for the might of the land of Biainili (and) for the pacification of enemy lands. The earth was a wilderness (?), nothing had been built there [previously]. [. . .] I settled there 6,600 warriors of the lands of Hatti and Tsupani [and] Supani. (Melikishvili 1960: no. 127)

21. Deportation was not a uniquely Urartian political strategy; it was also used by Neo-Assyrian and Neo-Babylonian rulers to mollify recalcitrant populations (Gallagher 1994; Oded 1979).

Altan Çilingiroğlu (1983) suggests that another large group of deportees from the Upper Euphrates area (the lands of Hatti and Melitea) were settled around Erebuni in 783 B.C.[22]

Through draconian programs of resettlement, the Urartian regime attempted to replace perceptions of a local landscape suffused with meanings derived from now-conquered polities with a sense of place entirely organized by the political landscape of Urartu. Urartian deportation programs were, in their fundamental character, quite similar to those pursued by various twentieth-century governments intent on dissolving bonds between people and place, such as the Ottoman Empire against Armenians or Stalinist Russia against Chechens. Like practices of site demolition, Urartian policies of deportation were meant not only to reorder the physical relationship between subjects and polity (for this could be accomplished without resettlement) but also to destroy the affective ties between people and place established by preexisting political communities.

The Urartian regime was thus interested not only in propagandizing through royal media, such as inscriptions, but also in more profoundly transforming the political subject through a brutal reordering of the cartography of memory achieved in both the destruction of place and the production of forgetting. Alongside this effort to manufacture the polity through a reformation of the subject, the Urartian regime attempted to shift the terms of political subjectivity through a reordering of the experience of landscape.

EXPERIENCE

Immediately following Argishti I's conquest of the Ararat plain (ca. 785 or 780 B.C.), the Urartian regime began a program of intensive building in the region that profoundly transformed the experience of the political landscape, drawing the region into the physical architectonics of the polity—the patterns of movement and action ordered by built physical form. The available archaeological data from the region allows us to discuss two primary axes of subject-polity architectonics: settlement location, and site topography. The location of settlements can be described in terms of two primary metric dimensions, elevation and position, which interact to define broad-scale geographic relations. Location is thus a regional measure, charting spaces occupied and unoccupied by built mani-

---

22. The practice of deportation appears to have continued in Urartu well into the eighth century B.C. An inscription of Rusa II declares that the king built new cities in *Ziukini,* an area that Arutyunian locates on the west coast of Lake Van, populating them with deportees from Hate, Mushki, and Halitu (Çilingiroğlu 1983: 323).

festations of political authority in order to define patterns of movement between them. Site topography provides a more localized account of flows into and out of settlements.

Thanks to the numerous archaeological and architectural surveys that have been conducted over the past half-century in southern Caucasia, we can create relatively comprehensive maps of the regional distribution of major pre-Urartian and Urartian period settlements (figs. 22 and 23).[23] However, by looking only at the distribution of the fortresses—not smaller sites, cemeteries, and the like—we should keep in mind that these maps describe only one aspect of a broader landscape. Location, in topographic terms, is a strikingly homogeneous trait of pre-Urartian fortresses of southern Caucasia. Of 32 known fortresses in the Ararat, Shirak, and Tsakahovit plains, all but four (Metsamor and Shamiram on the Ararat plain, Agin and Gusanagyukh on the Shirak plain) were located on the mountain slopes of the highlands rather than on the plain. The mean elevation above sea level of pre-Urartian fortresses of the Shirak and Tsakahovit plains was 2,054 meters; for the lower-lying Ararat plain region, the mean fortress elevation was 1,552 meters.

The rather peculiar locational bias of pre-Urartian fortresses, qua political centers, holds significant implications for the experiential architectonics of subject-regime relationships in pre-Urartian polities. The limited distribution of political centers of the early fortress-states defined a distinctive practical relationship between fortress and countryside, regime and subjects, centered on vertical movements up and down the mountain slopes. These centripetal ties linking subjects to fortresses were bolstered by the policing of territorial margins. Investigations in the Tsakahovit plain have described a set of small fortified outposts set high in the Pambakh mountains overlooking the northeastern edge of the plain (fig. 24). This set of five outposts extended from Berdidosh in the northwest to Aragatsi-berd in the southeast. Each fortress straddles a pathway into the Tsakahovit plain from the Vanadzor region to the north. Such conspicuous monitoring of inter-regional pathways suggests that, as early as the mid-second millennium B.C., the pre-Urartian complex polities of southern Caucasia were physically inscribing political territories (and hence the extent of sovereignty) through formal circumscription as well as through practices that established centripetal ties to political centers.

The most striking dimension of pre-Urartian fortresses is their inac-

23. Important archaeological surveys in the region include Adzhan, Gyuzalian, and Piotrovskii 1932; Areshian et al. 1977; Biscione 1994, forthcoming; Kalantar 1994; Piotrovskii and Gyuzalian 1933; Toramanyan 1942. See also Avetisyan, Badalyan, and Smith 2000.

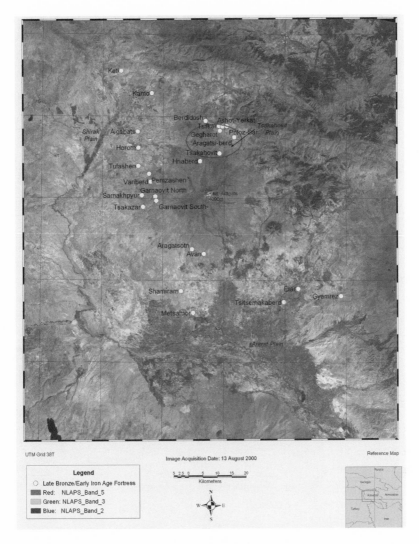

FIGURE 22. Late Bronze and Early Iron Age fortress sites in the region of Mt. Aragats superimposed on a Landsat 7 ETM+ image rendered in grayscale. (Source: Landsat 7 Earth Thematic Mapper.)

cessibility, created by a combination of topography and cyclopean masonry walls (fig. 25). The steep terrain surrounding pre-Urartian fortresses has generally been interpreted as a spatial response to the militarism of the era, suggesting that the mountain slopes surrounding the Ararat and Shirak plains provided security from raiding and conquest. This view is supported by a broad overview of changing settlement patterns in the re-

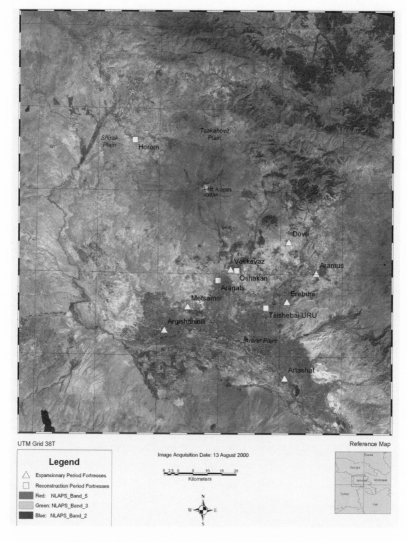

FIGURE 23. Urartian fortresses in the region of Mt. Aragats superimposed on a Landsat 7 ETM+ image rendered in grayscale. (Source: Landsat 7 Earth Thematic Mapper.)

gion. Extensive settlement on the Ararat and Shirak plains proper is in evidence only at times of considerable political stability such as that provided by the Pax Persica (546–331 B.C.) and Pax Armenia (190–69 B.C.). However, without an understanding of the material and ideological resources being contested, militarism is not a sufficient interpretation.

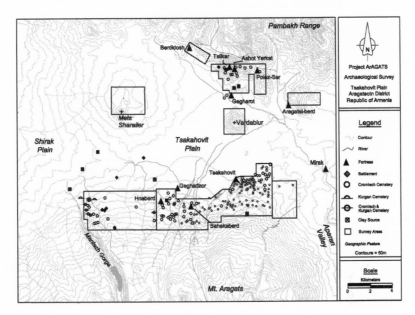

FIGURE 24. A map of known Late Bronze Age sites in the Tsakahovit Plain, Armenia. (Source: Project ArAGATS.)

The experience of the political landscape of the Ararat plain during the Urartian imperial period (from the conquest of the region in approximately 785 B.C. to the campaign of Sargon II in 714 B.C.) contrasted significantly with those of the preceding Late Bronze and Early Iron Age polities. Locations of political centers shifted dramatically from the mountain slopes toward the plain, indicating an intensification of direct political oversight. Whereas pre-Urartian fortresses clustered in the highlands surrounding the Ararat plain, Urartian sites were built at much lower elevations. The mean elevation of imperial period Urartian fortresses is only 1,137 meters (well below the 1,552 meter-mean for pre-Urartian fortresses in the Ararat plain), indicating a strong movement of the built apparatus of governance off of the mountain slopes and onto the plain.[24]

It is important to note that certain dynamics of architectural scale and political status were built into the physical landscape of the Urartian Ararat

24. Statistical comparison of elevations for Late Bronze/Early Iron Age and imperial period fortresses (using the Mann-Whitney test) indicates quite clearly that elevation differences between the sets are significant ($p < 0.001$). Interestingly, a similar transformation in settlement patterns has been documented in the Montaro Valley of Peru, with the incorporation of Late Horizon local polities into the Inka Empire (see Earle et al. 1980).

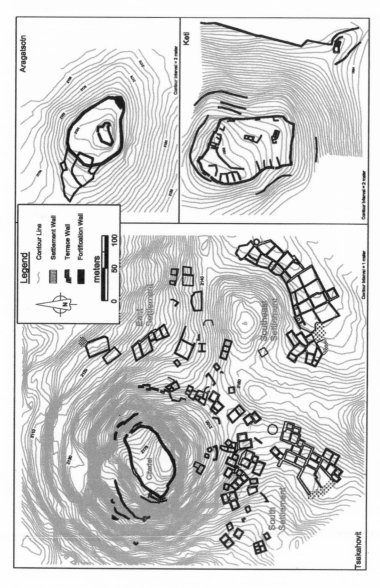

FIGURE 25. Architectural and topographic plans of three pre-Urartian fortified political centers from southern Caucasia. **a.** Tsakahovit. (Source: Project ArAGATS.) **b.** Aragatsotn. (Source: Project ArAGATS and Commission for the Preservation of Historical Monuments, Republic of Armenia.) **c.** Keti. (Source: Project ArAGATS and the Commission for the Preservation of Historical Monuments, Republic of Armenia.)

plain. The extent of the fortress of Argishtihinili, relative to its contemporaries, has led many to describe it as the primary economic center for the plain, whereas the fortress of Erebuni, thanks largely to its unique throne room, has been cast as the principal political center (the architecture of both sites is discussed in more detail in chapter 6). It is of note that the largest Urartian fortresses in the Ararat plain—Argishtihinili, Aramus, and Artashat—were located on the three primary routes into, and out of, the plain, suggesting an abiding concern for the exchange of resources with other regions of empire. Certainly the richness of the Ararat plain was peculiar to the Urartian empire as a whole, and its products were likely in demand in other provinces.

Yet the topography of imperial period fortress sites was not entirely unlike that of its predecessors. In general, Urartian sites throughout the kingdom perched atop hills and outcrops. However, the relief of Urartian sites overall is significantly less dramatic than that of its predecessors (fig. 26), indicating a significant diminishment in the importance of topographic relief to the production of the physical space of Urartian fortresses. Interestingly, the topography of Urartian fortresses in the Ararat plain contrasts markedly with contemporary sites in eastern Anatolia, such as Çavuştepe, where slopes boast a severity more akin to pre-Urartian political centers (Erzen 1988).

As a result of the distribution and topography of imperial period sites, Urartian fortresses were relatively accessible, compared to their Late Bronze/Early Iron Age antecedents. No topographical obstacles inhibited movement between the Urartian centers, and the terrain on which they were sited was significant but not forbidding. Political architectonics in the Ararat plain under the initial Urartian occupation were reoriented from the essentially vertical pattern established between subject and political apparatus defined by the pre-Urartian landscapes to a much more immediate spatial relationship. The political regime, no longer an aloof presence with its center of power high in the mountains, was a much more direct presence defining the polity through regularized surveillance of areas pivotal to the political economy.

The distribution of settlement in the Ararat and Shirak plains was only modestly altered in the reconstruction period of the seventh century B.C. that followed the political crises sparked by Sargon II's campaign (see fig 23).[25] The most drastic change in Urartian settlement during the

25. Although at least two imperial period settlements in the Ararat plain—Voskevaz and Argishtihinili—appear to have remained occupied into the seventh century B.C., we currently

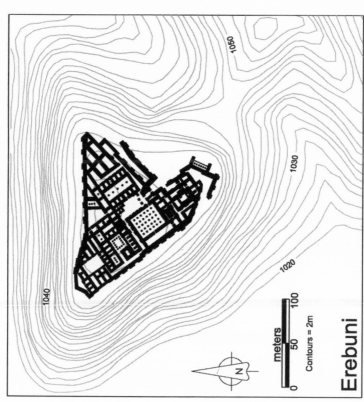

FIGURE 26. Architectural and topographic plans of two imperial period Urartian centers on the Ararat plain. **a.** Erebuni. **b.** (opposite) Argishtihinili. (Source: Project ArAGATS and the Institute of Archaeology and Ethnography, Republic of Armenia.)

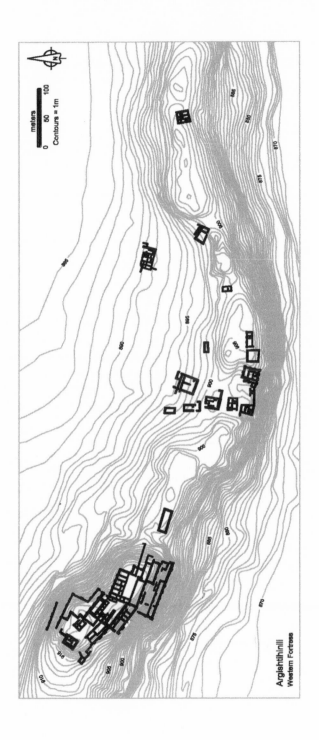

meters
0   50   100
Contours = 1m

N

Argishtihinili
Western Fortress

reconstruction period was the abandonment of the regional political center at Erebuni in favor of a new fortress—Teishebai URU (a.k.a. Teishebaini)—at Karmir-Blur, 7.3 kilometers to the west. The paucity of portable artifacts at Erebuni suggests that the site was systematically abandoned in the early seventh century B.C. Furthermore, the recovery of numerous objects at Karmir-Blur inscribed with the names of eighth century B.C. kings, some of which also bore the phrase "city of Erebuni," indicate that these "heirlooms," along with any political significance they may have carried, were transferred to the new fortress (see Melikishvili 1960: nos. 146, 147). This shift was accomplished in the first year of Rusa II's reign, contributing to a small decrease in the average elevation of settlements. The mean elevation of reconstruction period Urartian sites in the Ararat plain is only 977 meters, continuing the dramatic relocation of centers of political authority off the mountain slopes and onto the plain that was initiated during the preceding century.

The topography of reconstruction period sites, like that of the imperial period, departs from the affinity for high places with steep approaches on all sides that marked pre-Urartian political centers.[26] Teishebai URU (Karmir-Blur), though dramatically defended by the dramatic precipice of the Razdan River gorge on its northeastern side, sits atop very mildly sloping terrain to the west and south (fig. 27). Perhaps the most emblematic illustration of the Urartian regime's declining interest in places of dramatic relief is the small fortress at Aragats, which was constructed on a rise that, at its summit, reached only 10 meters above the surrounding plain (fig. 28; Avetisyan 2001). The massiveness of construction at Aragats in effect consumed the entire hill.

---

lack information regarding the disposition of the fortresses at Artashat, Dovri, Metsamor, Aramus, and Menuahinili and therefore cannot include them in the present discussion of the reconstruction period.

26. A statistical analysis of variance within the topography of each period indicates a highly significant ($p = 0.01$) decline in median surface grade—a measure of local topography—from the Early Iron Age through the reconstruction period. Median surface grade is a standardized measure based on slope calculations for eight profile lines radiating from the exterior walls of a fortress along the cardinal directions (for a more detailed discussion, see A. T. Smith 1996: 158). The overall median surface grade of the 17 pre-Urartian sites of the Ararat, Shirak, and Tsakahovit plains for which data are available is 30.8 percent, ranging from 42.5 percent (Keti) to 19.8 percent (Tufashen). The median surface grade for imperial period Urartian sites in the Ararat plain is 19.7 percent. This contrasts with contemporary Urartian fortresses in the Van region. For example, surface grade at Çavuştepe, southeast of Lake Van, ranges from 28 percent to 40 percent at the steepest locations. The median surface grade for reconstruction period sites on the Ararat plain is 16 percent, even lower than that of the imperial period. At Aragats, the approach to the site has a median grade of only 8 percent.

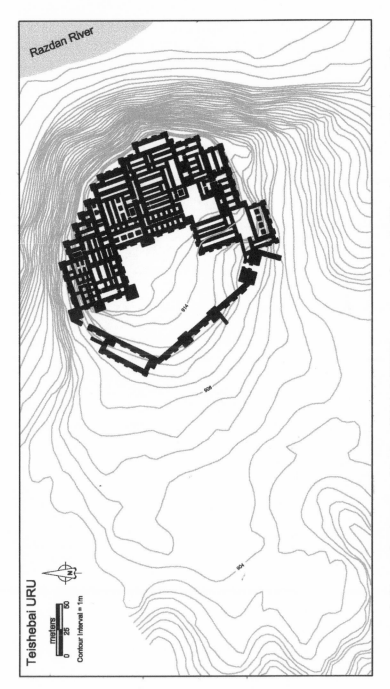

FIGURE 27. Architectural and topographic plan of Teishebai URU, a reconstruction period Urartian center on the Ararat plain. (Source: Institute of Archaeology and Ethnography, Republic of Armenia.)

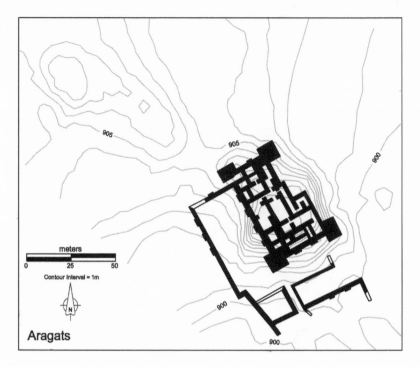

FIGURE 28. Architectural and topographic plan of the Urartian fortress at Aragats. (Source: Project ArAGATS.)

Local topography is significant for describing Urartian political practice in that defense of the polity and administration of its resources created opposing demands on sites. Whereas the former valued inaccessibility and distance, the latter prized openness and ease of transport. It appears that, by the reconstruction period, the demands placed on fortresses as elements in a regularized political economy of resource extraction and exchange had largely trumped the defensive utility offered by locating fortresses on sites of exaggerated topographic relief.

Authority and the Landscape of the Polity

Transformations in the political landscape wrought by the kings of Urartu detail the historically shifting practical relationships that mediated between the regime and its subjects. These transformations were directed toward producing the Urartian polity as a multidimensional political landscape: as an experiential landscape of regulated material and bodily flows, as a perceived landscape that built abiding commitments to place and destroyed rival senses of place, and as an imagined landscape summoned through the triumphal charisma of the king. However, the fit between these three dimensions of landscape was not always perfect. For example, the intensive redistributive economy of the kingdom demanded that Urartian centers in the Ararat plain be accessible, near critical trade routes, and close to centers of production and arable land. Yet traditions of political centers from the preceding 500 years privileged more inaccessible places built on sites of dramatic topographic relief. In the eighth century B.C., Argishti I appears to have balanced these demands in the Ararat plain by splitting the primary political center (Erebuni) from the primary economic center (Argishtihinili). Thus, the former was set atop a high rock outcrop in the foothills of the Gegham range, whereas the latter was set on a low mound adjacent to the Araxes river near the intersection of north-south and east-west trade routes. By the seventh century B.C., lingering concern with the evocative sense of place attached to pre-Urartian political communities seems to have waned in southern Caucasia: Teishebai URU, a consolidated political and economic center, was built on relatively flat ground on the west bank of the Razdan River. Such emergent gaps between the experiential, perceptual, and imaginative dimensions of landscape not only provide a critical impetus to reproduction but also constitute a conceptual locus for anthropological critique.

In a remarkable study of the constitution of the modern Thai nation-state, Thongchai Winichakul describes the changes in spatial practices, particularly the adoption of European traditions of cartographic representation, that transformed "premodern" Siam into the "geo-body" of the Thai nation: "[T]he term geo-body is used to signify that the object of this study is not merely space or territory. It is a component of the life of a nation. It is a source of pride, loyalty, love, passion, bias, hatred, reason, unreason" (1994: 17). The formation of the modern geo-body of Thailand was accomplished, he argues, through transformations in political practice accompanied by new understandings of space, predicated on the technology of modern cartography, which set aside indigenous understandings of the political landscape. Mapping became an instrument to concretize the national project that was so fervently desired by imperial regimes in London, Paris, and, ultimately, the sovereign authority of a new regime in Bangkok.

What this study suggests is that the formal demarcation of territorial boundaries alone does not make the polity. What establishes the polity are certain configurations of political practice established through the experience, perception, and imagination of landscapes that (1) regularize demands of regimes on subjects and (2) legitimate these demands in reference to both senses of place and descriptions of the proper world order. Policing boundaries and marking borders is one way to produce this landscape, as is reflected in the Egyptian frontier official's report that opened this chapter. But the establishment of formal boundaries is by no means the only element in the production of the landscape of the polity.

In the case of Urartu, the Ararat plain was incorporated into the landscape of the polity less by close policing of frontiers than by overlaying a physical landscape (one that tied subjects to regimes through regularized surveillance and a political economy focused almost exclusively on the governmental apparatus of the fortresses) with an imagined landscape framed by the tropes of the civilizing mission of the ruling regime. By directly attacking preexisting senses of place that might have provided an alternative vision of the political landscape, the Urartian regime produced a polity within which it held sovereign authority. Polities emerged not from formal demarcations but from practices that established and reproduced sovereign authority within relationships between subjects and regimes. But how can political subjectivity be defined within an archaeology of early complex polities?

Subjectivity is constituted not as an absolute but in the variable links of individuals to a sovereign political apparatus. Subjects are thus produced within the same practices that give rise to the landscape of the polity. Spatial ties are produced in order to establish individuals as subjects and to frame the parameters of political subjectivity:

> To be governed is to be at every operation, at every transaction, noted, registered, enrolled, taxed, stamped, measured, numbered, assessed, licensed, authorized, admonished, forbidden, reformed, corrected, punished. It is, under the pretext of public utility and in the name of the general interest, to be placed under contribution, trained, ransomed, exploited, monopolized, extorted, squeezed, mystified, robbed; then, at the slightest resistance, the first word of complaint, to be repressed, fined, despised, harassed, tracked, abused, clubbed, disarmed, choked, imprisoned, judged, condemned, shot, deported, sacrificed, sold, betrayed; and, to crown all, mocked, ridiculed, outraged, dishonoured. (Pierre-Joseph Proudhon, quoted in Pierson 1996: 58)

The French anarchist Proudhon captures in his list, spat out in phrasings that cannot conceal both his spite and his closeted admiration, what being a subject means: the surrender of some portion of will to another. In this surrender is the germ of authority that makes possible all other political relationships.

In the case of early complex polities, the subject has been discouragingly undertheorized. With the focus of investigation so resolutely centered on the State, subjectivity has been articulated only in the anthropologically thin terms of coercion by force of arms or consent through contractarian rewards. Neither of these are particularly robust models of what it was to be subject within an early complex society. Only at rare moments has the relation between authority and subject in early complex polities been more explicitly theorized as a relational problem for a political apparatus. James Scott suggests that "[t]he premodern state was, in many crucial respects, partially blind; it knew precious little about its subjects, their wealth, their landholdings and yields, their location, their very identity. It lacked anything like a detailed 'map' of its terrain and its people. It lacked, for the most part, a measure, a metric, that would allow it to 'translate' what it knew into a common standard necessary for a synoptic view. As a result, its interventions were often crude and self-defeating" (1998: 2).

Although it is difficult to deny that many episodes in which early complex polities dealt with their subjects lacked subtlety and at times backfired, is this the result of a lack of an adequate cartography of subjects? Were political authorities in early complex polities so ignorant of their subjects and, conversely, were subjects really so able to slip through the fingers of regimes? The political landscape produced in the Ararat plain, and indeed throughout the kingdom, would suggest that the Urartian kings had a well-developed understanding as to who their people were and what their terrain entailed. Indeed, the project of deportation in particular suggests that the Urartian regime had sophisticated knowledge of local senses of identity and their embeddedness in the surrounding environs.

The landscape of the polity—the spaces real, perceived, and imagined created by political communities as territories governed by a sovereign regime with authority over a discrete group of subjects—remains a central feature of contemporary politics. It is difficult to survey the history of the twentieth century without noticing that the politics of modern nations, like the politics of Urartu, lives in landscapes where the experiential order of spatial practices are part of a multidimensional reorientation of spatial practices that also attempt to transform the perceived links between people and place and the imagined sources of the landscape. The polity of Urartu was produced within sets of political practices that established it as a landscape in the physical, perceptual, and imaginative encounters between subjects and the sovereign regime. Yet this description begs the question of how regimes themselves articulate with landscape. It is to this topic that we now turn in the next chapter.

CHAPTER 5

Regimes

You ought to speak of other States in the plural number; not one of them is a city, but many cities, as they say in the game. For indeed any city, however small, is in fact divided into two, one the city of the poor, the other of the rich; these are at war with one another; and in either there are many smaller divisions, and you would be altogether beside the mark if you treated them all as a single State. But if you deal with them as many, and give the wealth or power or persons of the one to the others, you will always have a great many friends and not many enemies.

Socrates in Plato, *The Republic*

The urban precincts of Chichén Itzá, the first major post-Classic period (A.D. 925–1530) political center in the Yucatecan Maya lowlands, mark a significant departure from the Classic period Maya political landscape. Whereas the Classic period cities bore the personal imprint of individual rulers and dynasties in the form and aesthetics of major constructions (see chapter 3), Chichén Itzá bears the traces of a more plurally sited governmental apparatus (Stone 1999: 299). None of the pivotal events in the life of a ruler (birth, accession, death) that provided major narrative foci for Classic period monuments are recorded in early post-Classic hieroglyphs. Nor do we find at Chichén Itzá the sort of built dynastic genealogies that were recorded on the lintels at Yaxchilan (Schele and Mathews 1998: 234–39; Tate 1992). Much diminished are the pyramidal temple complexes (now represented only by the Castillo) that, during the Classic period, had so profoundly located authority in isolated rulers who mediated the terrain between earth and cosmos. Instead, what we find at Chichén Itzá is an urban fabric framed by novel architectural forms, such as the Temple of the Warriors, that instantiated political practices across multiple horizontal relationships rather than a singular vertical axis. Andrea Stone (1999: 314) has argued that this transformation in the physical space of the post-Classic urban landscape was a critical element in the expansion of Maya political practices to focus on institutions constituted as "common ground" on which factional shared governance took shape.[1] The lines of this factional competition and collaboration were likely drawn between rival "houses" whose membership was described in terms of kinship and lineage (Gillespie 2000: 467).

At the center of this transition from Classic to post-Classic political authority lies a profound reshuffling of the relationships constituting governing regimes, marked most conspicuously by the depersonalization of rule—what T. Patrick Culbert (1991: 327) has described as a separation of political offices from the charismatic personalities of office holders. This shift in Maya political regimes seems to have its origins in the terminal Classic when hieroglyphic inscriptions, once a guarded prerogative of the *k'ul ahau* (hereditary king or divine lord), began to record a proliferation of titles for other political leaders, such as the *sahal* (secondary lord). This expansion in the officially recognized members of Maya regimes is particularly pronounced at sites such as Copan, Yaxchilan, and Piedras Negras, where sahals erected their own monuments or appeared with the k'ul ahau on shared monuments. The built environment of Chichén Itzá can thus be understood as an integral part of a shifting alignment within Maya urban-centered political regimes that entailed both an ebbing of the power of the k'ul ahau and a broadening of authority sited in a number of factional leaders of powerful houses.[2]

The transformations in the cities of the Maya lowlands from the terminal to the post-Classic emphasize the close association between the constitution of the authority of political regimes and the form and aesthetics of urban political landscapes. A similarly close association between political authority and city spaces has been described for numerous other early complex polities. When provincial rulers were empowered under the

1. This is to suggest not that there was no factional competition during the Classic period but that the constitutive interests within this competition had changed rather significantly.

2. My thanks to Cynthia Robin for pointing out to me the profound spatial and political shifts that accompanied the Classic to post-Classic transition.

Zhou kings of early first millennium B.C. China, a new walled city, or *Guo,* was built for them as both their residence and the epicenter of regional sovereignty (Barnes 1999: 136). The relationship between political leadership and urban built environments has also been noted in the pre-Columbian Andes, where capital cities—such as those at Cuzco, Tiwanaku, and Chan Chan (and their secondary satellites)—served as seats of royal lineages and centers for cults of dead rulers. As Alan Kolata has pointed out, "The *raison d'être* of the Andean city was not fundamentally economic but political and ideological" (1997: 246–47). Given this close association between urban landscapes and political authority in early complex polities, it is rather peculiar that the dominant theorizations of urbanism—both ancient and modern—tend to downplay the significance of politics in shaping human settlements and tend to treat their form and aesthetics as epiphenomenal to sociopolitical transformation. The roots of this reigning counter-intuitive position extend well into the nineteenth century, when several studies established the city as a coherent and unified phenomenon of sociological investigation (most notably, Fustel de Coulanges' *The Ancient City.*) However, within studies of early complex polities, the (ancient) City emerged most profoundly in the early twentieth century, stripped of both its political sources and its unique spatiality, in the revolutionary protohistory of V. Gordon Childe.

In a now canonical account of the rise of early complex polities, Childe described the advent of the archaic State as an "urban revolution," a radical transformation in social relations represented most emblematically by the appearance of a novel form of settlement: the City (which, like the State, may be capitalized to emphasize its conceptual singularity). Sparked initially in southern Mesopotamia during the fourth and third millennia B.C., this second revolutionary moment in world history (following on the first, Neolithic, revolution)[3] was stimulated by the invention of new technologies—most notably bronze metallurgy and the wheel—and the attendant development of specialized commodity production dependent on long-distance exchanges of raw materials and finished products (Childe 1936: 159–62, 1946: 82–83). According to Childe, these economic transformations promoted the emergence of a priestly elite possessed of the authority to command both the material resources and labor of their com-

---

3. In Childe's narrative (1946: 41–44), the first revolution is the Neolithic, or agricultural, revolution in which settled human communities began to produce their own food through the domestication of staple crops, such as wheat and barley, and of herd animals, such as sheep and goats.

munities through their control over long-distance exchange networks. Monumental temples emerged as the central loci of this new economy, home to both the deity and the ruling "corporation of priests" that interpreted its will (1946: 96). The City arose around such monumental religious buildings as dramatic shifts in the means and relations of production generated densely populated settlements that provided labor for building programs, specialists for craft manufacture and processing, and a corpus of subsistence producers to feed the specialist and administrative classes (1936: 163–64, 1946: 84–87). From its primary bases in southern Mesopotamia, the Nile valley, and the Indus valley,[4] the urban revolution spread along trade routes to new centers in eastern Asia, Anatolia, the Levant, the Aegean, the Caucasus, and, ultimately, the heartland of Childe's Europe (1936: 195–200).[5]

Despite its titular prominence, Childe's urban revolution was not really about cities. True, cities emerge from Childe's theory as artifacts of class domination and loci of production and exchange within a commodity economy. But there is nothing about the form or aesthetics of the City, or any particular city, that either establishes this new socioeconomic constellation or drives its reproduction. The temple precinct plays a role in legitimating the demands of the elite on local communities, but this position is purely structural, not spatial. That is, some sort of religious structure is required as a focal point for ideological production, but how such structures were positioned within the urban landscape is thoroughly irrelevant to Childe's discussion.

The urban revolution was also not about revolution, at least not in the traditional sense of a rapid, radical overturning of political regimes (see the extended discussion in Trigger 1980). Although the corporation of priests establishes the institutional apparatus of government, this serves only as a tool for preserving economic privileges. The political holds no autonomy from the sectional interests of the dominant economic class, and thus "revolution only describes the cumulative attachment of a potent economic power to the preexisting sacred privileges that distinguished

---

4. Childe outlines simultaneous urban revolutions in Mesopotamia, Egypt, and India. However, the timing of the urbanization of the Indus valley now places it several centuries after urbanization in southern Mesopotamia (Kenoyer 1997), and the extent to which Egyptian settlements in the early dynastic and Old Kingdom periods can be regarded as "urban" is a matter of much dispute (Wenke 1997).

5. A similar recounting of the diffusion of the archaic State has been proposed by Andrew Sherratt (1997: esp. introduction and chap. 18), who locates the impetus to statehood across much of the Old World in the Uruk period transformations in southern Mesopotamia.

priests from the masses within Neolithic societies.[6] The City is thus not employed as a designation for human settlement at a certain spatial scale (as in the sociological tradition of Louis Wirth or Kingsley Davis).[7] Rather, the City denotes human settlement on a social evolutionary level, such that the relatively small, compact cities of classical Greece and the extensive rambling precincts of Teotihuacan are compressed into a single spatial category. In the familiar manner of spatial absolutism, Childe recast relations in space (such as variation in extent and density of settlements) as temporal relations within a metahistorical matrix.

Childe's account of the City thus rests on the twin foundations of a mechanical spatial absolutism, which links singular settlement form directly to a general world historical transformation, and a strict Marxist account of urban politics, where the interests of a dominant economic class exhaust the analysis of civil authority and governance. As such, Childe's account of the ancient City anticipated structural Marxist accounts of urban space and city politics forged in the wake of the pluralist challenge to elitist urban sociology in the early 1960s. Studies of urban politics by writers such as David Harvey and David Gordon have argued that historical stages in the development of capitalist economies determine urban spatial patterns, public policy, and even urban consciousness (Harvey 1973, 1985a, 1985b; Gordon 1977; Gordon, Edwards, and Reich 1982; see also Mollenkopf 1992: ch. 2, fn. 15). Neo-Marxist studies point out that capital, generally described as a unified interest, enjoys considerable privilege within political competition that leads to deeprooted and persistent reproduction of structural inequalities. Even as political coalitions compete on the margins over the exercise of power, the political agenda is in essence already determined by the enduring interests of those who control wealth.[8]

6. In later writings, Childe (1950: 3) suggests that he uses the term "revolution" to describe a dramatic rapid demographic explosion that resulted from "progressive change in the economic structure and social organization of communities."

7. In his seminal article "Urbanism as a Way of Life," Wirth (1938: 8) defines the City in physical and demographic terms as "a relatively large, dense, and permanent settlement of socially heterogeneous individuals." Similarly, Davis's (1965: 42) studies of urbanization utilize a strict demographic definition: "The difference between a rural village and an urban community is of course one of degree. . . . One convenient index of urbanization, for example, is the proportion of people living in places of 100,000 or more."

8. Claus Offe (1984) has extended the neo-Marxist position, arguing that the structure of urban governance must be understood in the twin imperatives to support elite economic interests (capital accumulation) and to achieve political legitimacy. Offe reasons that government compromises its legitimacy if it appears to be captured exclusively by

Although this strand of urban sociology has effectively highlighted the interdependence of political and economic elites in producing urban land-scapes, it tends to reductionism by suggesting that political and economic interests are always coterminous. As a result, structural Marxist accounts of the capitalist City have been slow to examine the mechanisms through which connections among elites are produced in relation to the links be-tween political elites and more grassroots sites of authority, such as neigh-borhood activists, local religious leaders, and prominent families. Fur-thermore, by assuming the analytical privileging of the economic over the political sphere, such ultimately functionalist accounts of the City, from Childe to Harvey, can never provide an understanding of how the imper-ative to economic privilege comes to dominate the operation of urban pol-itics. How is it that the ambitions of capital, in the capitalist city, or of com-modity exchange, in Childe's ancient City, come to dominate the political agenda? As John Mollenkopf has persuasively pointed out: "Such a stand-point begs the question of how these 'imperatives' are put in place and re-produced over time, which inevitably must be through the medium of pol-itics" (1992: 35). Fundamental to an account of ancient cities is not simply an analysis of the determining economic interests of elites but also the con-stitution of the political authority of more broadly sited regimes. These regimes are located at the confluence of horizontal ties among elites and vertical ties between dominant political authorities and grassroots social positions, such as kin leaders, neighborhood councils, and ward chiefs.

This chapter develops a critique of the ancient City and explores the production of urban landscapes as constitutive of the authority of polit-ical regimes. The primary geographic focus for this discussion is the city of Ur and its neighbors in southern Mesopotamia during the late third and early second millennia B.C.: places that, at least since Childe, have provided a prominent foundation for the ancient City.[9] The proper ob-ject of study, I suggest, is not the City but the political regimes that pro-duce urban landscapes as built environments and imagined places.[10] This shift in analytic perspective allows us to view cities, as Socrates advocated in Plato's *The Republic,* as multidimensional landscapes simultaneously

---

elite interests—hence the agenda of elites must be balanced with allocations of resources perceived to benefit wider segments of the population.

9. Surpassed only by the classical cities of Greece and Roman (see Finley 1977).

10. This perspective on cities as perceived and imagined places as well as experienced spaces reflects a pluralization of Lewis Mumford's (1937: 59) definition of the city as "a geographic plexus, an economic organization, an institutional process, a theater of social action, and an aesthetic symbol of collective unity."

riven by a multitude of social and political divisions yet pretending to co-herence through a highly politicized urban imaginary. Such a transfor-mation in the investigation of urban politics in early complex polities re-quires that we take a closer look at the history of the problem as it developed in southern Mesopotamia.

Urban Landscapes in Southern Mesopotamia

Southern Mesopotamia[11] occupies a privileged position in most histor-ical accounts of early cities because it appears to be the region that hosted the earliest appearance of large, densely settled, built environments (fig. 29).[12] The politics of southern Mesopotamia were, from a precociously early date, the politics of cities. As early as the Ubaid period (ca. 5900–4300 B.C.), several settlements in southern Mesopotamia are known to have reached sizes of 10 hectares or more (Pollock 1999: 45).[13] During the subsequent Uruk period (ca. 4300–3100 B.C.), a handful of settlements

11. The Greek toponym "Mesopotamia," translatable as "land between the rivers," came to denote the area between the Tigris and Euphrates rivers (modern Iraq and Syria) rather late in that region's history (Postgate 1992: 3; note the positioning of Mesopotamia between Armenia Major to the north and Babylonia to the south in Ptolemy's *Geography* [1991: 129]). Today the term refers more broadly to the entirety of the land between the Tigris and Eu-phrates from the foothills of the Taurus mountains in the north to the coastline of the Per-sian Gulf in the south—a region that encompasses the traditional territories of Assyria in the north and Babylonia in the south. Babylonia is generally applied to the southern reaches only following the rise of powerful dynasties at Babylon during the early second millen-nium B.C. Prior to that time, southern Mesopotamia is traditionally divided in half based on the dominant textual traditions in each area, with Sumer—or, perhaps more precisely, the portion of southern Mesopotamia where Sumerian was the dominant written language—in the south and Akkad in the north.

12. Jericho (located in the Jordan River valley) and Çatal Hüyük (on the Konya plain of south-central Turkey) have also been pointed to as potential early cities. Excavations at Jeri-cho revealed a stone tower and encircling wall in Neolithic I period levels (ca. 8500 B.C.); however, the site's claim to city status is belied by the limited extent of the site and the small number of houses within the walls (Holland 1997; Kenyon 1957). Çatal Hüyük's claim to be the birthplace of the city is predicated on the dense agglomeration of the site's architec-tural core; though quite large indeed, Çatal Hüyük nonetheless lacks many of the features common to sites defined as cities, such as street networks and fortifications (Mellart 1967). Recent work in northern Syria, at the remarkable site of Tell Hamoukar, threatens to dis-place both Jericho and Çatal Hüyük in efforts to define the first city (Gibson 2000). Although each of these early sites is of great interest because they challenge histories of urbanization, the theoretical stakes in defining any of these settlements as a city or not are exceedingly low and tend to merely reinstantiate the ancient City as an absolute space by privileging the ideal type over the specific forces producing a given urban landscape.

13. Dates for the Ubaid period represent the broadest extent of an era as defined in Roaf 1990: 51, 56.

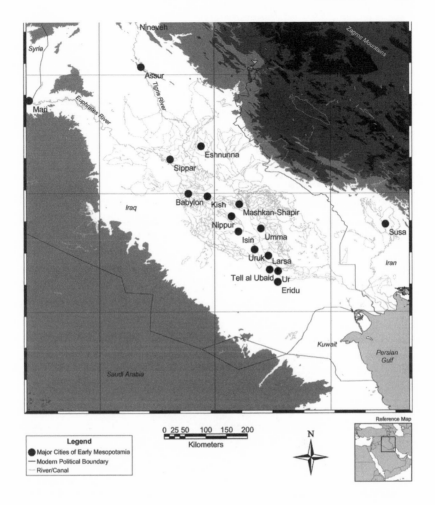

FIGURE 29. A map of Mesopotamia and its surroundings: major sites of the third and early second millennia B.C. (Map source: ESRI Data & Map CD.)

underwent explosive physical growth (Adams and Nissen 1972; Kouchoukos 1999; Pollock 2001).[14] Most notable among these early urban sites is the city of Uruk itself. At the beginning of the third millennium

14. On the dating and chronology of the Uruk period, see the synthetic analysis of the extant radiocarbon determinations in Wright and Rupley 2001. For an expansive discussion of various dimensions of society and politics during the Uruk period, see the papers in Rothman 2001.

B.C., Uruk's city wall enclosed an area of approximately 5.5 square kilometers, an unprecedented urban sprawl surpassed in extent only during the early first millennium A.D. with the growth of Rome, in the Old World, and Teotihuacan, in the New (Nissen 1988: 71). Beyond its substantial size, Uruk boasted a number of features that are now regarded as markers of distinctly urban built environments, including a circumscribing city wall, differentiated zones for residence, production, exchange, and religious and political institutions, and, to use Spiro Kostof's (1991: 40) phrase, a monumental architectural fabric that provided the city with a scalar sense of enormity. Foremost among the monumental buildings of Uruk period cities was the central temple complex, dedicated to the patron deity of the city. At the city of Eridu, archaeologists uncovered a series of superimposed temples that extended from the Ubaid period through the Uruk period, suggesting that a commitment to enduring sacred places did indeed play a role in the stability of early Mesopotamian urban sites and, quite likely, in the production of the earliest urban landscapes.

This process of urbanization had crested by the beginning of the Early Dynastic period in the early third millennium B.C. An array of urban political centers now parceled out the southern Mesopotamian alluvium into discrete urban-centered polities. However, an escalation in tensions among these polities, perhaps fueled by contests over territory, emerging local ideologies, or local resources, sparked an era of inter-city political competition that further promoted urbanization by creating unrest in the countryside and facilitated the full emergence of an authoritative political institution distinct from the temple, the palace (Stone 1995: 236).

This era of competing cities was brought to a close by Sargon (ca. 2300–2230 B.C.), originally of the city of Kish, who conquered the reigning regional polity that had been assembled by Lugalzagesi of Uruk as well as the remaining independent cities. (For an overview of the Sargonid era, see Kuhrt 1995: 44–55.) Sargon's conquests established direct imperial rule throughout a unified Mesopotamia and extended the influence of the Sargonid regime from the Taurus mountains to the Persian Gulf. Sargon founded a new city, Akkade, from which to rule his empire; however, the reign of the kings of Akkade was always turbulent. Despite a renaissance under Sargon's grandson, Naram-Sin, the empire collapsed just a few generations after it was established, and the Akkadian dynasty endured for only a short, rather inglorious period after the reign of Sharkallisharri. Authority was located once again in rival cities, establishing a fundamental historical oscillation in the region among eras dominated by local city authorities, periods of rival coalitions that fractured the geopolitical landscape, and times of regional coalescence. This tension was played out most emblematically in the centuries after the fall of the Sargonid empire, when an era of extreme political instability gave way to two major periods of coalescence under the Third Dynasty of Ur (ca. 2100–2000 B.C.) and the First Dynasty of Babylon (the Hammurabian dynasty, ca. 1800–1600 B.C.), interrupted by two centuries (ca. 2000–1800 B.C.) of coalition-based competition centered most prominently on the cities of Isin and Larsa.

Following the chaotic collapse of the Sargonid empire and the ensuing formation of small kingdoms at Lagash, Umma, Uruk, and a much reduced Akkade, Utu-hegal of Uruk installed Ur-Namma as his governor at Ur. (Hallo and Simpson 1971: 77 suggests that Ur-Namma may have been a close relative of Utu-hegal [see also Kuhrt 1995: 58–59].) Ur-Namma gradually asserted Ur's independence from Uruk by rebuilding the city wall and assuming the title "King of Ur." To further emphasize his break with Uruk, Ur-Namma sponsored a massive building program at the heart of the city, erecting the great terrace *(temenos)* and refurbishing the temples of the city deities Nanna and Ningal, including the construction of the massive ziggurat of Ur-Namma (Woolley 1939). Ur-Namma moved to increase his regional prominence by securing the approval of the priesthood of the city of Nippur, sponsoring the rebuilding of the temples of Enlil and Ninlil (head deities in the Sumerian pantheon). These efforts won him the title most coveted by southern Mesopotamian rulers with regional ambitions: "King of Sumer and Akkad." In matching political reality to this title, Ur-Namma appears to have resorted to violence at least once, defeating rivals at Lagash and thus establishing a politically unified southern Mesopotamia centered at Ur.

During the reigns of Ur-Namma's son (Shulgi) and grandsons (Amar-Sin and Shu-Sin), the kings of Ur presided over a highly centralized governmental apparatus that asserted political authority over various facets of daily life, intensifying demands on production and setting forth the first "law code" (see Roth 1995a: 13–23). But under his great grandson Ibbi-Sin, Ur's empire gradually sagged, some argue, under the weight of its highly centralized, intensely politicized economy (Civil 1987; Steinkeller 1987). In attempting to forestall the crisis, Ibbi-Sin empowered Ishbi-Erra, the commander of his northern troops, to rule a largely independent province incorporating both Isin and the symbolically charged center at Nippur. This move seems to have kindled a broad fracturing of the Ur III dominion and the empowerment of a number of rival polities centered on several prominent cities. When Ur was sacked by the Elamites, and Ibbi-Sin was carried into exile, there was considerable competition in both the south and the north that also involved polities beyond the immediate reaches of southern Mesopotamia, such as Elam. For a time, the kings of Isin emerged as the most prominent regional power, ruling Ur through appointed governors. Only with the emergence of powerful regimes at Uruk, Larsa, and Babylon in the nineteenth century B.C. was Isin's prominence checked and internecine wars ultimately quelled under the auspices of a new fully regional hegemon, Babylon. Under Sumulael, Babylon achieved ascendancy in the north, followed several generations later by Hammurabi's consolidation of the southern polities.

From even this necessarily brief overview of early Mesopotamian history, it should be clear that regional politics were first and foremost urban politics. The centrality of urban worlds to early Mesopotamian politics has been succinctly described by Marc Van de Mieroop: "The Mesopotamian known to us today was a citizen, a resident of one of these ancient towns. The art and literature we admire was created by these citizens, the bureaucracies we study were urban bureaucracies, and the politicians and military leaders we know were living in cities" (1997: 1–2). Yet the role that urban form and aesthetics played in constituting Mesopotamian political authority through their regulation of the experience, perception, and imagination of spatial practices remains largely unexamined, in large part because of the manner in which the intersection of space and politics has been theorized over the past century and a half of archaeological research in the region. Thus, before proceeding to an examination of Mesopotamian urban landscapes, it is important that we frame the problem in reference to contemporary theory.

Theorizing Urbanism in Early Southern Mesopotamia

The antiquarian vision of ancient Mesopotamian cities was not simply particularistic, as it has often been described, but self-consciously exegetical. The explication of a handful of canonical texts—the Bible, the Homeric epics, *The History* of Herodotus, Xenophon's *Persian Expedition,* Arrian's *History of Alexander*—provided the intellectual lodestar for antiquarian explorations of cities throughout southwest Asia. The impetus to research within this antiquarian passion for exegesis lay in the possibility that archaeological remains and secondary epigraphic sources could provide a political history that might be correlated with biblical and classical accounts of events (Dever 1997: 315). The political history sought from these sources was less analytical than recitative, focusing on the places and periodizations of kingly reigns. The most profound effect of this limited focus on political history was to vest accounts of early Mesopotamian politics in a series of great rulers and their cities—from Sennacherib and Semiramis of Nineveh to Nebuchadnezzar and Hammurabi of Babylon—a stance reflected in Paul Emile Botta's invitation to Austen Henry Layard to visit the excavations at the neo-Assyrian site of Khorsabad (Dur Sharrukin): "[A]s for you my dear Sir you are not a true lover of Semiramis if you do not come here [to Khorsabad] to superintend my work" (quoted in Larsen 1996: 26).

The antiquarian focus on cities as the dominions of individual rulers left political life woefully undertheorized but the space of the ancient City highly overdetermined. That is, the cities of ancient Mesopotamia were understood to flow directly from the construction programs of kings without any account of either the processes that translated royal policy into built edifices or the array of interests that might have intruded on the imposition of kingly authority on physical space. In this sense, the antiquarian cities of ancient Mesopotamia were quintessential subjectivist spaces, read as the physical imprint of the majesty of ancient rulers. For the antiquarian, urbanism was of interest only insofar as the monumental fabric of urban construction provided a sense of aesthetic achievement. The antiquarian imagination was more explicitly architectural than it was urban, as Layard himself described it: "Visions of palaces underground, of gigantic monsters, of sculptured figures, and endless inscriptions floated before me. After formulating plan after plan for removing the earth and extricating these treasures, I fancied myself wandering in a maze of chambers from which I could find no outlet. Then again, all was reburied, and I was standing on a grass-covered mound" (1970: 25). Indeed, early antiquarians such as Layard and Botta did not really excavate cities (or even buildings) as complete structures, focusing instead on tunnels and deep trenches targeted toward recovering the sculpted reliefs that decorated the walls of Assyrian palaces (Liverani 1997: 89).

This antiquarian subjectivism was challenged by twentieth-century absolutist accounts of urban space that arose out of new commitments in archaeological thought to social evolutionary patterns over historical personalities and to materialist explanation over exegesis. The two writers most profoundly responsible for shaping this turn in the context of early Mesopotamia were Childe and, later, Robert McC. Adams.

The antiquarian accounts of ancient Near Eastern cities had derived much of their intellectual force (and immense popularity with the general public) from their evocative articulation of specific places with memorialized events. Thus, cities such as Babylon and Jericho, Mycenae and

Troy were compelling in their very specificity as the ground once tread by biblical figures or Achaean heroes (a trope that continues to define biblical archaeology and endures to a lesser extent in classical archaeology). In proposing the urban revolution as a singular historical process rooted in a vision of the essential sameness of the City, Childe was explicitly writing against this genealogical, subjectivist tradition in antiquarian historiography. In order to compress the particular genealogies of the antiquarians into a single global history, Childe advanced a mechanical absolutist ontology of space in human social transformations by defining a set of universal spatial forms attendant to each stage of economic development. The Neolithic revolution in food production brought with it the Village and the urban revolution brought the City, a singular, absolute space. All parts of Childe's City—palaces and temples, marketplaces and production areas, residences of the wealthy and residences of the poor—map directly onto each other as partners in the simple structural play of elite reproduction. None hold analytical autonomy from the others, and the spatiality of the urban fabric holds no determinative significance for the interplay of elements within a historical drama centered on class domination.

The spatial absolutism of Childe's ancient City rests on his description of urban politics, a view derived in large part from contemporary understandings of the Mesopotamian remains. Having ceded determinative priority to the relations of production within an emergent system of commodity production and exchange, regulation of the City was vested entirely in a priestly elite that preserved its class privilege through the apparatus of government and gained legitimacy from its proximity to the sacred. Power arose from economic privilege, legitimacy from ritual chicanery. Childe's City thus fit quite comfortably within contemporary elitist urban sociology, where politics was rendered as coterminous with the interests of a relatively homogeneous, structurally stable, social class (see Mills 1956; Schulze 1958; Schulze and Blumberg 1957).

Childe's focus on the structural determination of City form within the economic interests of a privileged elite remained profoundly influential in histories of early urbanism, even as archaeological and epigraphic discoveries created severe problems for his overall account of the urban revolution in Mesopotamia.[15] The most influential discussion of the ancient

---

15. A number of highly influential studies of the City during the 1960s and 1970s reflect the continuing influence of Childe on accounts of urban origins. Jane Jacobs's (1969) fanciful description of the origins of the city in the Near Eastern obsidian trade owes a pro-

City after Childe was undoubtedly Adams's comparative study, *The Evolution of Urban Society*. Adams's brief, yet extraordinarily erudite, comparative investigation of early complex societies in Mesopotamia and Mesoamerica was framed as an attempt to come to terms with Childe's legacy—to systematize the urban revolution as a historical process rooted firmly in the transformation of the sociopolitical order attendant to the formation of the State. That is, by whittling away the broadly cultural features of Childe's second revolution—such as writing, representational art, and transport technology—Adams attempted to establish the City on firmer sociological foundations. In choosing to compare two cases, the cities of Mesopotamia and central Mexico, Adams explicitly sought to control for any possible "genetic" links between urban social worlds and thus arrive at an account of the "lawful" regularities that operated within the urban revolution (1966: 20). Like Childe, Adams was intent on establishing the City as an absolute space, freed from the obfuscatory particularities that might highlight variability.

However, unlike Childe, who formulated the political as a superstructural transformation driven by altering economic relations rooted in long-distance commodity exchange, Adams focused on the articulation of a broader set of sociopolitical locations with profound material commitments to local spheres of production (ibid., 14). According to Adams, urban society emerged as a product of the intertwined social evolution-

---

found debt to Childe in the central position accorded the rise of raw material (obsidian) extraction, processing, and long-distance commodity exchange in stimulating urban development. However, rather than arising out of mechanisms for processing raw materials brought in from afar, New Obsidian, Jacobs's imagined original urban community, develops thanks to its proximity to, and domination of, obsidian sources. Alternatively, Paul Wheatley's (1971) *The Pivot of the Four Quarters* (1971) highlights the central role played by ritual elites in the making of urban sites out of ceremonial centers. Whereas Jacobs inherited Childe's account of exchange economies, Wheatley developed Childe's sociological vision, elaborating the structural centrality of religious officials to urban development. More recently, Aidan Southall's (1998: 27–29) survey of the City follows Childe in tracking its evolution across a progressive series of modes of production. Within Southall's version of the "Asiatic" productive mode, the "pristine" City, such as arose on the alluvium of southern Mesopotamia, developed out of the economic exploitation of holy shrines by priests (Childe's "corporation"). These priests became progressively divorced from subsistence and material production thanks to the offerings brought to the emergent towns from the surrounding countryside. The City thus developed as a symbolic expression of the supernatural, made possible by agrarian surpluses and an increasing long-distance trade in luxury goods supported by the new class of elites. Politics is largely absent from Jacobs's economic model, and in Southall's account it is an entirely secondary development marked, at least in the Mesopotamian context, only by the eventual cleaving of a secular institution of kingship from the temple priesthood.

ary transformations of transhumant foraging into settled agriculture, kin affiliations into stratified social classes, and specialized priesthood into administrative apparatus. The City formed around these transformations as the spatial locus of political authority and the center of a densely settled population. Adams's account of urban political life is richer and more plural than that embraced by Childe insofar as political economy, social class, and governmental administration do not map directly onto each other but provide potentially discrete arenas of activity.[16] However, the interests of each of the dominant sociopolitical locations in material production, particularly subsistence economies, do create a certain singularity that allow Adams to root broad cultural transformations, in the last instance, in local ecological conditions.

Adams describes the rise of the City within an "ecological mosaic" of distinct yet interdependent zones of specialized subsistence production. Cereal agriculture, garden and orchard horticulture, and pastoral and piscatory production all combined in southern Mesopotamia to create a complex human ecology that was mediated by the central institutions around which the City grew. The City thus emerged as a spatial nexus, mediating exchanges among occupationally specialized subsistence producers located in the rural hinterlands. The ecology of the southern alluvium during the third and second millennia B.C. does indeed suggest a region that hosted a variety of ecological zones supporting a broad variety of subsistence resources. Areas of irrigated cultivation were intersected by numerous natural waterways and built canals. Together with marshes and grazing land, southern Mesopotamia supported a diverse array of fishes, waterfowl, domestic livestock, wild fauna, arboreal produce, and cereal crops.

By highlighting the role of the surrounding environs in promoting the emergence of the City, Adams shifted the basic ontology of archaeological studies of early urbanism from Childe's mechanical absolutism, where spatial form arose purely in relation to evolutionary stages, to a more organic absolutism, where local environments play, if not a fully determinative role, then a critical enabling one in giving rise to the City.[17] Since the publication of *The Evolution of Urban Society*, much of Adams's work

16. See also pluralist critiques of elitists in urban sociology, especially Dahl 1961; Polsby 1960; Sayre and Kaufman 1960.

17. The organicist vision of ancient Mesopotamian history remains prominent in the literature today (e.g., Algaze 2001b; Redman 1999). But compare Adams 2000 for a statement on the current neglect of ecological issues in social evolutionary accounts of early complex societies.

can be read as an effort to problematize organic absolutism insofar as it allows for simple deterministic readings of relationships between humans and environments. Adams's peerless regional surveys in southern Mesopotamia and continuing work with remote sensing data have brought to light the human role in producing key elements of the regional ecology: the extensive canal systems that allowed cultivation on the alluvium; the salinization of the soils around intensively occupied sites created by intensive demands on productivity by political regimes beginning in the third millennium; the degradation of topsoils promoted by overgrazing; and the effects on piscatory resources caused by varying demands on the region's hydrology (Adams 1965, 1978; Adams and Nissen 1972; Redman 1999: 127–39).

Very early in the development of complex societies in southern Mesopotamia, the local environment had already been profoundly shaped by political policies and decisions. As Nicholas Kouchoukos has persuasively argued, "At the nexus of environment and human society is landscape — the concepts, perceptions, and patterns of behavior with which human beings interact with their environment and on which they depend for stable and sustainable social life" (1999: 177). Given the thoroughly created nature of the Mesopotamian countryside, attempts to root the spatiality of early Mesopotamian polities in the local ecology demand that we then provide an account of the canals, divided lands, pasturage set asides, and so forth within a description of contemporary politics and the interests that drove such a far-reaching program of landscape transformation.

Despite Adams's call for studies of early urbanism to focus on "societal variables," the social world is rendered in *The Evolution of Urban Society* in terms of "paradigmatic models" that aid an evolutionary project but cloak the spatiality of social practices. Thus, Adams was able to suggest that urbanism was less important than "social stratification and the institutionalization of political authority" (1966: 9–10). However, such a position begs the question as to how stratification and authority were constituted and reproduced over time, which inevitably must be, at least in part, through the instruments provided by urban landscapes as settlement forms and as imagined places. That is, spatial practices of urbanism and political practices of authority are not separable. Adams has himself moved in this direction in recent years. In a study of Sasanian period irrigation systems in southern Mesopotamia, Adams has called attention to the prominent role that geopolitical concerns and factional rivalries within the political apparatus drove a massive reorganization of the physical landscape of town and countryside in a web of causal interdependencies laid out quite elegantly by the Sasanian king Khusro Anushirwan (A.D. 531–579): "Royal power rests upon the army, and army upon money, and money upon the land-tax *(kharaj),* and the land-tax upon agriculture, and agriculture upon just administration, and just administration upon the integrity of government officials, and the integrity of government officials upon the reliability of the vizier, and the pinnacle of all of these is the vigilance of the king in resisting his own inclinations, and his capability so to guide them that he rules them and they do not rule him" (Altheim and Altheim-Stiehl 1954: 46; cited in Adams n.d.; see also Rubin 1995).[18] Organic absolutism demands that we cut this web of interdependencies once we reach agriculture and the articulation of human systems with natural environments. However, as Anushirwan well knew, and as Adams's work since *The Evolution of Urban Society* has demonstrated quite profoundly, the manufactured nature of local environments demands that forays into the constitution of authority understand the places of subsistence production as elements of political landscapes, as thoroughly manufactured as cities and towns.

In the decades since Adams's seminal work, the ancient Mesopotamian City has received considerably more attention within general studies of urban history than as an object of study within investigations of early complex polities. This move away from the City reflects, in part, a shift in archaeological practice generated out of Adams's articulation of the City with its surrounding countryside and environmental context. This move out of cities has produced a number of important investigations of small villages and shifted the geographic parameters within which the study of early complex polities in southwest Asia now treads (see, particularly, the papers in Schwartz and Falconer 1994). However, this shift in the archaeological gaze did not bring with it a retheorization of the spatiality of political life. Hence, as archaeologists have come to focus on urban landscapes once again, they have done so with the traditional absolutist ontology largely intact. Thus, the literature on urbanism in early complex societies remains steeped in either mechanical evolutionism, where regularities in historical transformation bring with them a specific set of spatial forms, or various strains of ecological possibilism. (For a prominent example of the former, see Marcus 1983; for examples of the latter, see Algaze 2001b; Sanders, Parsons, and Santley 1979; Redman 1999.)

Van de Mieroop has provided the most systematic recent examination

---

18. My thanks to Robert McC. Adams for sharing with me his recent work on Sasanian irrigation systems and the imperial authority of Khusro Anushirwan.

of the ancient Mesopotamian City. He opens his study by cogently pointing out that "[t]here is no such thing as *the* Mesopotamian city, as each
one of the hundreds that existed had its own peculiarities, and to gener-
alize from one of them would be misleading" (1997: 5). Yet the thrust of
Van de Mieroop's study is indeed the articulation of a model for the Mesopotamian City that is broadly enduring across the 3,000 years or so of
early regional history and is valid for both Babylonia and Assyria—an endeavor predicated on an absolute spatial ontology. Like the formulations
of the City proposed by Childe and by Adams, Van de Mieroop argues
that shifts in political regimes are epiphenomenal to the understanding of
urbanism (ibid., 8). That is, rulers came and went but the essential formal
character of the Mesopotamian City remained unaltered. Van de Mieroop
succinctly summarizes his foundational absolutism, writing "The state was
built around the city" rather than the city being built around the shifting
relationships constituting regimes. The result is a timeless urban space to
which authorities fit themselves rather than a dynamically changing urban
landscape that is produced, in large measure, by political practices.

Van de Mieroop holds to this image of a stable, unchanging urbanism
despite the fact that Mesopotamian topologic terminology seems not to
have recognized a general class of settlements apart from specific configurations of political sovereignty. Because of the specificity of this linguistic argument, it is worth quoting Van de Mieroop at length. Within native Mesopotamian terminology,

> An enormous number of settlements were referred to with the Sumerian and
> Akkadian terms we translate as 'city': *uru* [Sumerian] and *ālum* [Akkadian].
> The Akkadian term was used for anything from the metropolis of Babylon in
> the sixth century to a farmstead with seven inhabitants in the area of Harran
> in the seventh century. It was used for the entire city of Nineveh as well as for
> a section of it. . . . The translation 'city' is thus misleading since we classify
> settlements by size, and reserve the term for larger ones. . . . The lack of differ-
> entiation among settlements seems to reflect a perception that all of them were
> equivalent and sovereign communities. (Ibid., 10)

This observation would seem to suggest not only the instability of the
City (as defined in modern intellectual traditions) as a conceptual apparatus for understanding Mesopotamian urbanism but also that politics
was central to Mesopotamian understandings of urban life. In turning
now to a discussion of the elements of Mesopotamian urban landscapes
as seen from the southern city of Ur, the central analytical problem is not
how we might strip away the specificity of an urban environment to reveal the conceptual heart of the City but rather how these very particu-

larities of physical form, environmental aesthetics, and representation were instrumental in constituting the authority of urban political regimes.

Urbanism and Regime

The following examination of southern Mesopotamian regimes and urban political landscapes focuses on the Ur III and early Old Babylonian periods (ca. 2119–1880 B.C.).[19] This 240-year period witnessed a profound set of transformations in both the formal organization of governing regimes and the cultural representation of political order. Ideally, we would be able to tease apart the lines of these transformations in a diachronic account of changing political landscapes. Unfortunately, the archaeological record at present does not provide such historical sensitivity. Instead, I want to pursue here the parallel sociological argument that southern Mesopotamian cities were produced as multidimensional landscapes that assembled the power and legitimacy of complex, multi-sited regimes. These regimes were assembled out of the horizontal ties among key elites located in pivotal institutions (such as temple and palace) and vertical ties to more grassroots social positions diffused across the city (such as neighborhoods and wards).

IMAGINATION

The Epic of Gilgamesh—a story set in the early third millennium B.C.— opens with a profoundly visual tableau.[20] The titular hero, king of the city of Uruk, stands atop the magnificent city walls that he brought forth from divinely established foundations, inspecting the brickwork and reflecting on the monumentality of his achievement:

> [Gilgamesh] built the rampart of Uruk-the-sheepfold,
>   of holy Eanna, the sacred storehouse.
> See its wall like a strand of wool,

19. The dates used here for the Ur III and early Old Babylonian periods reflect the conventional dating of the period from the ascendancy of Utu-hegal of Uruk to the enthronement of Sumulael at Babylon (Kuhrt 1995: 63, 79). The early Old Babylonian period is also referred to as the Isin-Larsa period.

20. Despite its third millennium setting, the Gilgamesh epic appears to have been formally assembled late in the second millennium B.C. Copies of the standard version are known primarily from copies unearthed in the library at Nineveh of the Assyrian king Assurbanipal (Moran 1995: 2330).

view its parapet that none could copy!
Take the stairway of a bygone era,
    draw near to Eanna, seat of Ishtar the goddess,
that no later king could ever copy.
Climb Uruk's wall and walk back and forth!
    Survey its foundations, examine the brickwork!
Were its bricks not fired in an oven?
    Did the Seven Sages not lay its foundations?
[A square mile is] city, [a square mile] date-grove, a square mile is clay-pit,
    half a square mile the temple of Ishtar: [three square miles] and one half
    is Uruk's expanse.

(George 1999: I.i)

Gilgamesh's tectonic authority—his ability to produce the urban fabric of Uruk on a place sanctified (and thus legitimized) by the gods—established a close association in the Mesopotamian literary tradition between the built environment of cities and the exercise of political authority. Construction of the city was rendered as the primordial political act, a mustering in stone, brick, or mud of the authority of governing regimes. Alternatively, the destruction of the city, as described in the *Lamentation over the Destruction of Sumer and Ur,* represented an unmitigated catastrophe, an unleashing of the forces of chaos on the land and an upending of the civil order:

The city of Ur is a great charging aurochs, confident in its own strength,
It is the Primeval city of Lordship and Kingship, built on sacred ground,
(the gods) An, Enlil, Enki, and Nimah decided its fate. . . .
Revolt descended upon the land, something that no one had ever known,
Something unseen, which had no name, something that could not be
    fathomed
The lands were confused in their fear,
The god of that city [Ur] turned away, its shepherd vanished.

(Michalowski 1989: ln. 52–55, 65–68)

The end of the Third Dynasty of Ur (conventionally dated to 2004 B.C.; Kuhrt 1995: 63) was represented in the lamentation not simply as a conquest, as an overturning of one regime for another, but as a crisis of both the power of kings and the legitimacy of a sociopolitical order that was given by the gods.

However, city building in Mesopotamia was not simply a transformation of the megalomania of kings into mud brick as theorized by the antiquarians—a raw display of power to command capital and labor. Rather, the legitimacy of city building was highly problematic, as evidenced by the handful of attempts by Mesopotamian rulers to challenge the primacy of established cities through the construction of political capitals outside of the historically dominant urban places. Disembedded capitals—centers of regional political administration removed from established urban locales—are relatively rare in Mesopotamian history.[21] Although the neo-Assyrian kings of the early first millennium B.C. founded a number of new sites from which to rule the empire, these were generally framed as elaborations of existing cities. Thus, when Tukulti-Ninurta moved his court out of Aššur, the historic center of Assyrian politics, and across the Tigris River, he represented the move not as the founding of a new city but as an expansion of the harbor area of the capital. Only with Sargon II's movement of the peripatetic neo-Assyrian court to the newly constructed site of Khorsabad do we find boasts of founding a new city couched within an imperialist trope of a civilizing mission (a trope reminiscent of the expansion of the Urartian polity described in chapter 4).[22] However, the Khorsabad project was not an enduring one; Sargon's successor, Sennacherib, abandoned it unfinished and relocated the royal court once again: this time to the ancient city of Nineveh, which he substantially rebuilt in what might be regarded as the earliest program of "Haussmannization."[23]

Prior to the wanderings of the Assyrian royal court, we know of only two good examples of disembedded capitals in Mesopotamia: Dur-Kurigalzu, established by a Kassite king of Babylon in the early fourteenth century B.C., and Akkade, capital of the Sargonid empire. Kurigalzu did not frame the founding of the new metropolis that bore his name as the establishment of a new city. No records survive marking the foundation

21. I use the term "disembedded" in the sense used in Joffe 1998 and Van de Mieroop 1997: 59–61, as places limited to governmental administration, removed from more plural sites of social and economic activity. For example, Washington, D.C. began as a capital disembedded from traditional places of social and economic activity. This is generally not the sense of the term employed in the debates during the 1970s over the location of Monte Alban within the Oaxacan polity (see Blanton 1976: 257–58).

22. Sargon II's framing of the founding of Khorsabad as civilizing a wilderness is a rather unique account of the political production of new urban spaces within the Mesopotamian tradition. Indeed, it fits more comfortably within traditional Urartian descriptions of construction than Mesopotamian imperial discourses. Such an observation raises the possibility that Sargon II borrowed Urartian imperial rhetoric in attempting to legitimize the founding of Khorsabad. Although such a suggestion demands closer argumentation, it does provide a means for challenging traditional descriptions of Urartu as a receiver of Mesopotamian political wisdom by raising the possibility that Assyrian rulers borrowed from the Urartian kings.

23. For an orientation to Baron Haussmann's reconstruction of Paris, see Olsen 1986: 35–57.

of the city (modern Aqar Quf) in toto. Instead, what has come down to us is a series of texts marking the foundations of individual buildings rather than the city as a whole. The implication, as cogently argued by Van de Mieroop (1997: 53), is that the site was conceptualized not as a new city to rival the established urban landscapes but simply as an assemblage of new institutional loci that were not afforded the same sense of place that marked Mesopotamian cities. Alternatively, after Sargon ascended to the throne at Kish in the twenty-fourth century B.C. and began assembling the cities of southern Mesopotamia under his suzerainty, he established a wholly new city, Akkade, from which he and his successors would rule a united Mesopotamia. Although Akkade may not have been completely abandoned for many centuries, after the unraveling of the Sargonid empire it never again played a significant role in Mesopotamian geopolitics except as a spectral place within the Mesopotamian political imagination. A document dating to the early first millennium B.C. chastised Sargon for founding a new city and thus incurring the wrath of Marduk, patron deity of Babylon:[24] "He (Sargon) dug up earth from the clay pits of Babylon and built a replica of Babylon next to Akkade. Because of the wrong he committed, the great lord Marduk became angry and wiped out his people by famine. From east to west there was a rebellion against him, and he was afflicted with insomnia" (Van de Mieroop 1997: 59).

Van de Mieroop quite reasonably suggests that the rarity of new city construction in Mesopotamia, and the reluctance of kings to boast of large-scale construction projects as programs in city founding, reflects a culturally embedded suspicion of such enterprises: "In my opinion, the reluctance of Mesopotamian kings to boast about their city building was grounded in the general attitude towards the merits of such an enterprise. Founding a new city was considered to be an act of *hubris*" (ibid.). It would seem, then, that the imagined urban landscape of Mesopotamian cities such as Ur, Uruk, Aššur, and Nineveh was predicated on a sense of their coherence historically, as a limited set of privileged potential loci of political authority, a restricted group of enduring places that *legitimately* constituted the political landscape of the region.[25] Thus, one critical element

24. An anachronism, because only several centuries after Akkade had fallen did Babylon (and its patron deity Marduk) emerge as a powerful regional political center.

25. As Geoff Emberling (1995) has pointed out, during the third millennium B.C., cities appear to have provided critical loci for the production of personal identity in southern Mesopotamia. What remains unclear are the forces driving the creation of such imagined communities. It seems arguable that efforts to stake personal identity in cities were part of the broader efforts of regimes to reimagine cities as coherent locales of authority.

FIGURE 30. A carved stone relief from the palace of Sennacherib at Nineveh (ca. 704–681 B.C.): city models offered to the King of Assyria. (From Place 1867.)

in establishing a sense of the Mesopotamian urban imaginary was this sense of political authority rooted in a clear genealogy of place.

The spatial coherence of cities within Mesopotamian traditions was most emblematically represented in the palace reliefs of the neo-Assyrian kings, where models representing conquered cities are shown being brought before the king as tribute (fig. 30). The use of a similar iconography of cities can be found in northern Mesopotamia at least as early as the mid-second millennium B.C. when a number of Middle Assyrian period seal impressions depict city gates as synecdochic images of the city form (see Andrae 1938). By the second half of the second millennium B.C., the City was already a potent representational form, with pronounced architectural elements able to represent myriad sociopolitical relations. Such iconic use presumes, and reproduces, an imagined understanding of the coherence of the City as a spatially unified place.

Earlier iconographic and textual sources from southern Mesopotamia also bear on the production of the urban landscapes of the region as coherent and singular places within the contemporary political imagination. A foundational element in this imagined city is the link between the gods

FIGURE 31. Upper section of the stela of the law code of
Hammurabi, ca. 1792–1750 B.C. Recovered at Susa; carved
of diorite. (Photo courtesy of Réunion des Musées
Nationaux/Art Resource. Photo: Hervé Lewandowki.)

as the creators of divine places and the king as builder of urban monuments.
Atop the stela of Hammurabi, a 2.25-meter-high slab of black diorite bearing the most famous extant exemplar of that monarch's famous law code,
is an image of the king approaching the sun god, Shamash (fig. 31). Shamash
is seated on a throne whose rabbetted facade suggests the entryway to a
temple. In his right hand, the deity holds out to Hammurabi the tools of
a builder—the rod-measure and the rope-measure. (On the identification
of these items as measuring devices, see Frankfort 1996: 104; Jacobsen 1987:
4.) This image has been interpreted in a number of ways: as an offering to

FIGURE 32. Fragment from the Stela of Ur-Namma. (Photo courtesy of the University of Pennsylvania Museum Photo Archives.)

the king of the emblems of sovereignty; as a communication of the laws from the king to the god (or vice versa); or most persuasively, as a bequeathing to Hammurabi of the tools used to set buildings, particularly temples, on divinely ordained foundations (Roth 1995b: 22–23).

The prologue to the laws that begins immediately below the image provides what can easily be read as a literary elaboration of the pictorial representation: "When the august god Anu, king of the Anunnaku deities, and the god Enlil, lord of heaven and earth, who determines the destinies of the land, allotted supreme power over all peoples to the god Marduk, the firstborn son of the god Ea . . . named the city of Babylon with its august name and made it supreme within the regions of the world, and established for him within it eternal kingship whose foundations are fixed as heaven and earth" (Roth 1995a: 76). The text expands on the architectural description of kingship provided by the surmounting image, establishing a coterminous relationship between the foundations of the city and the foundations of the institution of kingship—both of which were set within a broader vision of the cosmos. The king as architect establishes the foundations of the city according to the proper order set forth by the god(s). Thus the critical features of urban landscapes were produced within the Mesopotamian imaginary not simply as coherent spaces fixed in the cosmos but also as places legitimately called forth only by authority of the king.

A similar image can be seen in the second register of the Stela of Ur-Namma from Ur (fig. 32), the best preserved of the five vertically superimposed panels carved in relief on a 3 × 1.5 meter stone block (Canby 2001: catalog no. 14, pl. 31; Woolley 1974: 75–81, pl. 41). In this scene, Ur-Namma stands before the moon god Nanna (city deity of Ur) who sits on a throne with a rabbetted facade to indicate the built form of the temple. In his left hand, drawn close to the body, Nanna holds a battle-axe, while in his right hand, extended toward the king, he holds the rod and rope. In the register directly below the seated deity, a king carries a pick and other building tools, suggesting the image of an architect as well as of a royal builder. Thus the images atop the Hammurabi and Ur-Namma stelae provide a pictorial elaboration of Gilgamesh's architectural description of the construction of the walls of Uruk. In each, the king establishes the foundations of the monumental fabric of the city, including fortification walls and temples, instantiating a profound metaphor of the king as architect and builder and of the urban landscape as a manifestation of a divine order.

Three sociological locations were thus brought together within explicitly political forms of representational discourse: the city, as a coherent and unified place; the temple and city deity, as metonyms for the cosmos; and the king, as the authoritative producer of the urban landscape. What is at work, I suggest, is a highly complex effort to shape the spatial imaginary, one that simultaneously reduces the city to the temple — an enduring trope in Mesopotamian literature since the Early Dynastic period and perhaps earlier—lending it coherence in reference to the cosmos, and inserting the king as architect and builder, transforming an explicitly political landscape into an enduring transcendent one.[26]

The imagined landscape as depicted in political representations describes a coherent place of legitimate authority produced directly and exclusively by the king at the behest of the gods. The imagination of Mesopotamian urban landscapes carried in these representations raises two questions of the physical built environment. First, were Mesopotamian cities also perceived and experienced as fully coherent places, or were the images discussed above obscuring something more complex about the urban landscape? Second, was the king in fact the exclusive source of the urban landscape? If, as one might suspect, the centrality of the king in Mesopotamian media was more of a claim to legitimacy than a description of the real operation of power, then what were the alternative sites of authority producing these urban landscapes and how did they relate to each other and to the king?

PERCEPTION

In addition to the powerful images of kings calling forth cities under the auspices of the deities, the built aesthetics of Mesopotamian cities themselves reinforced a sense of their cohesiveness as real sociopolitical units. The most powerful built feature promoting the perception of urban coherence was undoubtedly the city wall, a construction that differentiated the city from the countryside and produced a clear marking of inclusion and exclusion. By the mid-third millennium B.C., city walls had come to be charged with high political symbolism as markers of both political autonomy and subservience. When the Ur III kings established their rule over a city, whether by coercion or negotiated capitulation, one of their first acts was to raze its walls (Postgate 1992: 252). The purpose of this activity was twofold. In a pragmatic sense, the destruction of the walls removed a serious obstacle to control and, if necessary, reconquest of the vanquished city. But the razing of city walls also symbolized the sub-

---

26. The trope of ruler as architect was revived by Stalin, who was accorded the title "Chief Architect and Builder of Our Socialist Motherland" (Tinniswood 1998: 156).

servience of the conquered urban place to an external rule. A city without a wall was a city ruled by a nonlocal regime. As the Third Dynasty of Ur began to unravel, losing the ability to project its authority into various subject cities, local rulers began to assert their independence from the regional hegemon. Their defection was powerfully evoked in their reconstruction of city walls. Such a move, of course, revived an instrument of self-defense. But, more powerfully, the erection of walls symbolized the reassertion of political independence.

The most architecturally imposing features of many southern Mesopotamian urban landscapes were the large city temple complexes. The temenos at Ur (fig. 33), a walled rectangular enclosure roughly 8.4 hectares in area set slightly northwest of center, housed the various components of the temple of the city god Nanna (including the massive ziggurat constructed by Ur-Namma), the *Enunmakh* (storehouse of Ningal, wife of Nanna), and the *Giparu* (home of the Entu priestesses).[27] C. Leonard Woolley's (1939: pl. 84, 85) reconstruction of the Ur-Namma ziggurat suggests three large inset rectangular platforms, the lowest measuring 62.5 × 43 meters, that together soared above the city to a height of 25 meters. As a result, the temenos complex certainly would have dominated the skyline as well as the plan of the city. And as home to the patron deity of Ur, it was the focus of both ritual practice and massive investments of material resources and labor.

However, the temenos was not the only religious space within the city. Two other general forms of temple spaces are visible in the archaeological record from the early Old Babylonian period. First, two rather sizeable temples were uncovered outside of the temenos: the Enki Temple along the southeast city wall, and the Nin-giz-zida Temple along the southwest city wall. Both temples were built, or perhaps extensively restored, by Rim-Sin I of Larsa during that city's ascendancy in the nineteenth century B.C. Rim-Sin's investment in the temples outside of the temenos compared to Ur-Namma's investments in the Nanna complex suggest somewhat different priorities in the urban landscape of Ur in relation to religious spaces, perhaps a consequence of the kings' very different relationships to the major institutions of the city and their role in securing royal authority.

Second, a number of what Woolley referred to as "wayside chapels" were also discovered in the residential precincts in the Ur III and early

27. In a reexamination of the Ur temenos, Jean Claude Margueron (1982: 156–67) suggests that Woolley's reconstruction of the temenos overestimates its size considerably by including the *Ehursag* (see discussion of palaces below) within the temenos wall. If Margueron is correct that the Ehursag lay outside the temenos proper, then the area of the Nanna complex would have been approximately 5.4 hectares.

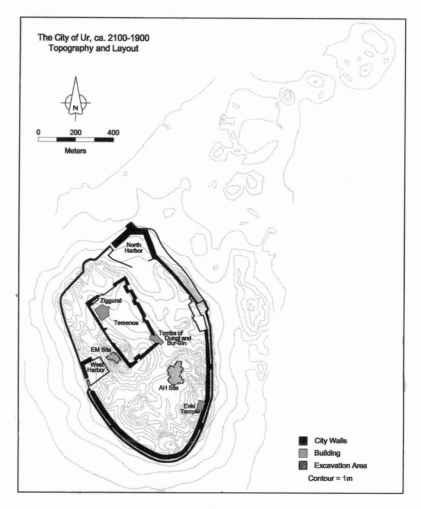

FIGURE 33. Architectural plan of Ur during the Third Dynasty and early Old Babylonian periods with modern topographic contours. Inset: detail plan of the temenos. (Redrawn after Woolley 1974.)

Old Babylonian period levels. The excavators defined wayside chapels based on three general architectural criteria: a prominent position in relation to the street, reveals in the doorway,[28] and an interior surface above street level (Woolley and Mallowan 1976: 30). Dedicated to minor deities, these wayside chapels, such as that located on the corner of Church and

---

28. Reveals are right-angled niched set-backs in the walls flanking a doorway.

Straight Lanes in the AH site, appear to have been open to a broader range of the population (compared to the rather closed areas of the large temples) and supported less by royal sponsorship or institutionally held assets than by local contributions. Indeed, these small chapels seem to have also served as neighborhood registrars; numerous tablets bearing primarily on real estate transactions (presumably those of local families) were found in the sanctuary of I Church Lane (room 4). The nesting of ritual spaces from the city temples of the temenos, to the temples of other deities, to the small wayside chapels likely had a powerful effect in forging a sacral community within the fragmented experiential landscape of Ur.[29]

In contrast to the generally well documented temple complexes, few features of early Mesopotamian cities are as elusive as the administrative and residential complexes of city rulers, usually described by the shorthand term "palace" (See Winter 1993). Although kings dominated the political landscape of early southern Mesopotamia, the spaces of the royal household and administrative bureaucracy are not particularly well known; only a handful of buildings in southern Mesopotamia have been established as centers of the royal apparatus. The most thoroughly explored palaces in Mesopotamia tend to be those from the north—the palace of Zimri-Lin at Mari or of the neo-Assyrian kings at Kar-Tukulti-Ninurta, Nineveh, Nimrud, and Khorsabad. However, a few buildings traditionally identified as palaces are known from southern cities. Two possible palatial structures are known from the Ur III period—the *Ehursag* at Ur, and the Palace of the Rulers at Eshnunna (Tell Asmar). Two additional palaces

29. Although excavations at other contemporary sites have not yielded the same diversity of temple spaces, evidence of major temple complexes are known from numerous other major cities. As at Ur, the main temple complexes at a number of sites throughout southern Mesopotamia were similarly located off-center. The ziggurats at Isin and Sippar were sited close to the edge of the city, whereas major temple complexes at Nippur were placed in a rather offset position in relation to the surrounding city walls. The variability in major temple siting in relation to the overall shape of the city suggests a certain independence of temple location from the overall shape of the urban environment. Furthermore, though it has been suggested that the corners of the temenos enclosure at Ur were oriented to the cardinal directions (a less than convincing description; see Lampl 1968: 14), no clearly identifiable regularities articulating temple geometry to a specific cosmological map of the world have been found to underlie religious architecture (as has been posited for numerous other cases, such as the Classic period Maya, where temple architecture appears to set forth in brick or stone a map of the cosmos [Ashmore 1986, 1989]). Instead, what seems to have been most determinative of temple locations within southern Mesopotamian urban environments was the commitment to historical precedent. As a result, any account of the temple within the urban landscape must be framed within local histories of the relationship between the growth of a city generally and episodes of temple reconstruction rather than in relation to enduring cosmological diagrams.

date to the Old Babylonian period—the palace of Nur-Adad at Larsa, and the palace of Sin-kashid at Uruk.[30]

There is little regularity in the positioning of these palaces within the urban landscape. However, the palaces at Larsa and Eshnunna and the Ehursag at Ur were constructed immediately adjacent to temple complexes. The palace of the rulers at Eshnunna, rebuilt during the Ur III period by the son of the local governor, adjoined a large shrine dedicated to Shu-Sin, a deified king of Ur's Third Dynasty (Oates 1986: 50). The Ehursag at Ur was constructed either within the temenos (according to Woolley [Woolley and Mallowan 1976]) or just outside its southern wall (as suggested by Margueron [1982]). However, the palace of Sin-kashid was far removed from the central religious complex at Uruk. It does seem reasonable to suggest that among the considerations in the siting of royal complexes was proximity to sacred places. But it is worth noting that there is a considerable difference in framing proximal relations between palace and the main temple of the city deity (such as at politically independent Ur) and between palace and a temple to a deified sovereign from another city (such as at politically dependent Eshnunna). Though united by a theme of sacrality, they are separated by very different political content.

What emerges from an account of the nested temple complexes distributed throughout Ur and the site of the palace just inside, or adjacent to, the temenos is a sense of the interdigitation of these institutions and their architectural locations with the city as a whole during the Ur III and early Old Babylonian periods. In the case of Ur, the question is not whether such edifices occupied the exact center of the city or were oriented to the cardinal points, but rather how these sites served to evoke a sense of the fundamental unity of the city as a sociopolitical formation. By establishing the Ehursag in the shadow of the temple of Nanna, the Ur III kings forwarded a practical sense of the spatial continuity of the urban regime in its two most critical sites, the temple and the palace. However, though the aesthetics of inclusion and exclusion may have promoted a sense of the coherence of the urban landscape, the ambiguous formal and locational links between palace and temple do suggest the possibility that the king was not the sole source of the urban landscape, as presented in contemporary media, but rather that a series of interlinked sites of elite and grass-

30. For a thorough discussion of the problems in identifying palaces in Mesopotamia generally and the questions that linger over the particular edifices noted above, see Margueron 1982. See also Flannery 1998 for a discussion of several other third millennium B.C. buildings that might be interpreted as palaces.

roots authority played prominent roles in creating the urban landscape and preserving a more broadly sited urban political authority.

EXPERIENCE

Few sites in southern Mesopotamia provide us with a synchronic view of the built environment of sufficient breadth to detail the central elements shaping the experience of an urban landscape. It is in part for this reason that general portraits of the Mesopotamian City have attempted to assemble an idealized account of urban form from numerous discrete cases drawn from some 3,000 years of regional history. Of course, the compression necessary to build the ideal type precludes posing the very questions that I would like to explore in this portion of the study: what role did political authorities play in forging the experience of a city; where were they sited; and how did their use of physical form contribute to the reproduction of existing constellations of political authority?

One of the few sites boasting broad exposures across a variety of contemporary urban areas is the city of Ur, once located on one of the many watercourses associated with the Euphrates River (see fig. 33). Excavations at the site from 1922 to 1934 led by Woolley recorded a variety of urban features dating from the early third millennium B.C. through the Neo-Babylonian period of the mid-first millennium B.C. But the most extensive early remains at the site date to the era of the Third Dynasty and the early Old Babylonian period. This roughly 200-year span provides us with a view of the city both at its height, as the center of a unified southern Mesopotamian polity forged by the Ur III kings, and under the control of an external ruling authority after the Third Dynasty's collapse ushered in a new era of regional geopolitical competition.

Early urban sites in southern Mesopotamia, particularly those like Ur that were occupied for centuries or millennia, present a distinctive profile on the modern landscape of southern Mesopotamia. Built primarily of mud brick, as structures fell into disrepair or were razed to make room for new constructions, the settlement was steadily raised above the level of the surrounding terrain. It was this process of settlement construction and rebuilding that formed the raised earthen mounds, or tells, that give southern Mesopotamian cities their distinctive appearance and preserve their rich temporal depth in stratigraphically superimposed levels. The topography of the mound at Ur, a tell that extends some 1.1 kilometers along an elongated northwest-southeast axis and 0.75 kilometer northeast-southwest, is irregular, cleft by depressions that follow the pathways of ancient canals and roads and surmounted by hillocks that mark areas of intensive construction. The tell today rises 20 meters above the surrounding plain at its maximum height, and at sunrise and sunset in the late third and early second millennia B.C. the central mound would have thrown a long shadow, both literally and figuratively, across the surrounding territory.

Circumscribing Ur, and indeed most early urban sites in southern Mesopotamia, was a large wall built of mud brick and punctuated by a series of gates that afforded access between city and countryside.[31] At some sites, the city wall was complemented by a moat linked to the river or canal that ran by the city. The city wall of Ur, as constructed by Ur-Namma during the Ur III period, was a massive construction 3.13 kilometers long and ranging from 25 to 34 meters in width. The wall circumscribed an area of approximately 64 hectares.[32] Mesopotamian city walls evidence great formal variability in overall morphology and design. The walls of Ur trace an irregular curvilinear shape, following the topography of the mound. Other city walls employed angular turns that created more rectilinear forms. A Kassite period map of Nippur represents the walls of that city as a series of irregular sized segments joined at sharp angles rather than the rounded joins of the curvilinear walls of Ur (fig. 34).[33] Only during the first millennium B.C. do the rather opportunistic lines of Mesopotamian city walls seem to have been replaced by more carpentered segments and right angles, like at Babylon, where the walls formed a great bracket around the Arakhtum (fig. 35).[34] Although city walls certainly provided defense during times of war, they also served a broader role in everyday urban experience. City gates allowed for the regulation of urban traffic by limiting entry into the city to a handful of easily surveilled choke points. The walls of Ur formed the point of contact between the city precincts and the external world of the countryside.

Complementing the city walls and gates in regulating movement into

31. In northern Mesopotamia there was a tradition of using a series of nested rings of city walls, but the southern Mesopotamian preference seems to have been for a single line of massive wall (though the Ur temenos enclosure wall can be interpreted as an inner citadel wall).

32. In comparison, the city wall of Uruk, at its fullest extent, was 9.5 kilometers long, supported by approximately 1,000 semicircular bastions. The much more modest circumscribing wall at Ischali was only 8 meters thick and considerably smaller than the constructions at either Ur or Uruk.

33. It is interesting to note that the walls of Mashkan-shapir included both a rectilinear eastern segment and a curvilinear southern extent (Stone 1995).

34. The Arakhtum was either a branch of the Euphrates or perhaps the river itself prior to a shift in its course.

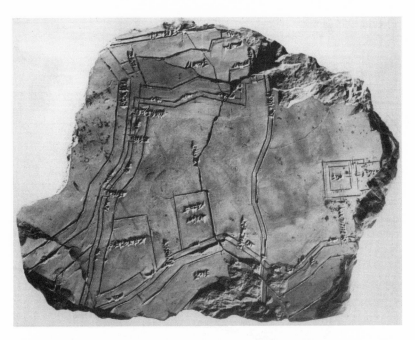

FIGURE 34. Kassite period map of Nippur inscribed on clay tablet. (Photo courtesy of the Oriental Institute, University of Chicago.)

and out of Ur were the *kārums,* or city harbors.[35] Woolley recorded two riverine kārums at Ur, one located on the northeastern edge of the city and one on the western flank. Both of these were linked by canals into regional exchange networks, thus providing a critical terminus for the movement of goods both within the polity and beyond. Similarly, Elizabeth Stone and Paul Zimansky (1994) recorded two contemporaneous kārums at Mashkan-shapir, the second capital of the kingdom of Larsa during the eighteenth and nineteenth centuries B.C., but these were more centrally located along the city's northeastern canal. Unfortunately, these harbor areas are known primarily from the textual records of shipping and exchange rather than from excavations, and so their architec-

35. *Kārum* is the Akkadian word used to denote the harbor areas or mooring places for riverine traffic within southern cities. In the Old Assyrian period, the term kārum was also applied more generally to places of exchange and merchant activity, such as the land-locked kārum at the Anatolian site of Kanesh. The kārum at Kanesh appears to have been part of an Assyrian trading colony established at the site during the early second millennium B.C. (Veenhof 1980).

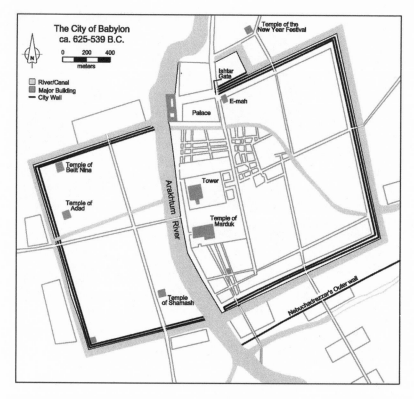

FIGURE 35. City plan of Babylon during the Neo-Babylonian period of the mid-first millennium B.C. (Redrawn after Oates 1986.)

tural features and configuration are largely unknown to us. However, it is important to note that the kārums at Ur were not open, easily accessible ports like Piraeus (the port of Athens), but rather closely guarded loci of merchant activity bound into the city within the walls themselves. They were thus constructed as areas of potentially strict regulation and control.

The topography of early southern Mesopotamian city sites is usually quite uneven, extending across several mounds separated by intervening depressions. These depressions have been shown in several contexts to mark the courses of urban canals that provided cities with drinking water and linked the kārums to riverine trade networks. But, as Stone has pointed out, the canals also served to fragment the urban landscape on a macro-level by parsing cities into discrete parts. Stone and Zimansky's surface survey at the Old Babylonian site of Mashkan-shapir suggests some functional differentiation of urban zones, with a focus on ceramic production in the northeast, a concentrated religious quarter in the southwest, and areas for administrative facilities, metal working, and lapidary production at discrete points in between. Stone concludes that "[t]he decisive role played by these watercourses in the structuring of Mesopotamian cities cannot be overemphasized. Mesopotamian cities were physically divided into different sectors by these canals: religious, administrative . . . and residential-artisanal sections" (1995: 239). The Gilgamesh epic also describes the city of Uruk as a divided landscape composed of four discrete parts: the city, the orchards, the claypits, and the open ground of the Ishtar Temple. Three-quarters of the city was thus rendered as open or largely unbuilt ground. Furthermore, the Kassite period map of Nippur also records a park in the southwest corner of the city.

Ur is more topographically unbroken than many cities in southern Mesopotamia. By the latter half of the third millennium B.C., the city was focused on one major mound with a series of smaller mounds extending to the northeast. But the central mound does have a broad fissure separating the southeastern lobe of the tell, perhaps reminiscent of the marks left by channels at Mashkan-shapir. However, it is not clear at present exactly how the internal lines of canals extended through the city. Although perhaps segmented by canals and major thoroughfares, the most spatially divisive construction at Ur was the temenos, whose walls effectively blocked traffic across the central portion of the northwestern lobe of the mound. As an analogy we might think of Central Park in New York City if it were to be elevated onto a terrace and enclosed by a massive wall. The effect of such an enclosure on the experience of the physical landscape of the city was to restrict movements across the city to a small number of routes.

To summarize the general physical outline of the city, though Ur appears to have been less conspicuously segmented than some contemporaries (such as Mashkan-shapir), there are indications that movement through it was restricted by the division of urban space into discrete fragments. This pattern of macro-level fragmentation—so conspicuously different from the imagined and perceived landscapes of coherence and continuity—was strongly amplified in the micro-level architectonics of the residential districts at the site. Because much of the argument of this chapter hinges on contrasting the experiential fragmentation of southern Mesopotamian cities with their representation as highly coherent places, it is important that we examine the micro-level architectonics of one site with broad Ur III and early Old Babylonian period exposures in some detail.

In only a very few southern Mesopotamian sites have archaeologists excavated significant portions of urban residential districts.[36] Excavations at Nippur uncovered several household complexes dating to the Old Babylonian period (see Stone 1987). Fortunately, more extensive exposures of residential districts were excavated by Woolley in the early Old Babylonian levels at Ur. Woolley conducted excavations in two residential areas, the EM site (fig. 36) adjacent to the southwest wall of the temenos, and the AH site (fig. 37), 150 meters southeast of the temenos. Of these, the AH site is the most extensive, covering 0.7 hectare and comprising, according to the excavators, more than 45 discrete residences on eight streets.

The most immediate formal characteristic of Ur's residential precincts is the density of the built space. For the AH site, the ratio of built to unbuilt space within the residential district is approximately 8.5/1, indicating a very dense architectural site. The primary transit axes through the AH site—Church Lane, Paternoster Row, Store Street, Broad Street, and Straight Street (named by Woolley after the byways of Oxford)—radiate from a central intersection (Carfax). Linked to these main routes were a number of dead-end streets and alleyways, such as Straight Street and Bazaar Alley.

Given the densely built nature of the site, most of the circulation beyond the streets and alleys was sequestered within residential spaces.[37] By abstracting the spaces and pathways within this dense network of roads and interior rooms (fig. 38), we can develop a sense of regularities in the structuring of the circulation through residential quarters of the city.[38] Four basic circulatory patterns are most readily identifiable from the built space of the AH site. The most straightforward circulatory pattern was a simple linear row of rooms that extended in series from the street (for example, I Niche Lane or VII Paternoster Row). Two variants on courtyard-centered circulatory patterns were also found throughout the site. In the radial pattern, rooms were entered via a single central court but did not have connections among themselves (for example, VIII Straight Street or XI Church Lane). In contrast, the nested pattern utilized a number of courtyards and multiple entrances to create more open, symmetric architectonics (for example, XI or IV/IVA Paternoster Row). A final circulatory pattern visible in the AH site organized room space in a highly

36. For an excellent overview of Mesopotamian residential neighborhoods during the Old Babylonian period, see Keith 1999.

37. One important exception, however, were the small "wayside" chapels, such as the two just off Carfax (I Paternoster Row and I Church Lane).

38. For a discussion of the techniques behind permeability graphs, see Hillier and Hanson 1984 or Markus 1993.

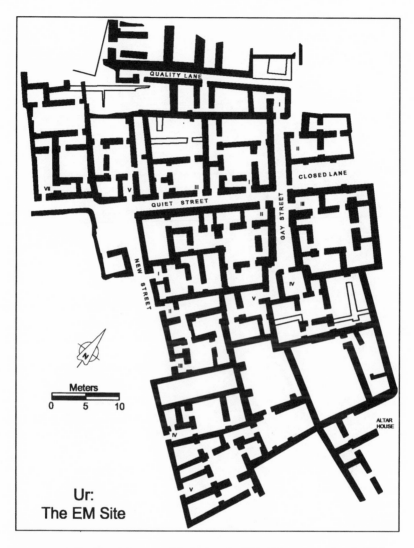

FIGURE 36. Residential areas at Ur during the Third Dynasty and early Old Babylonian periods: the EM Site. Roman numerals designate house numbers. (Redrawn after Woolley 1974; Woolley and Mallowan 1976.)

symmetric arrangement such that most or all spaces were equally proximal to the others (for example, I Paternoster Row).

The circulation diagrams suggest that room complexes were often formed by conjoining two previously independent blocks and, conversely, by subdividing residences. For example, I/IA Broad Street appears

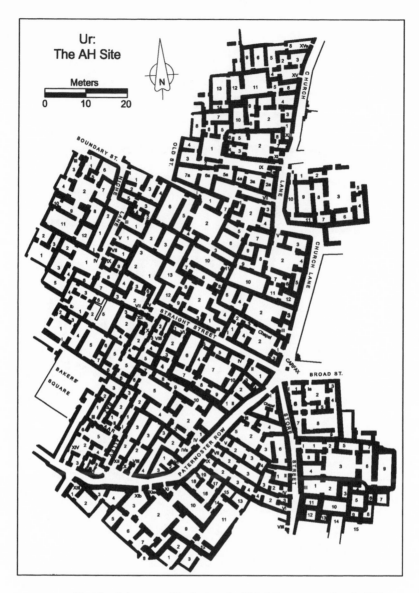

FIGURE 37. Residential areas at Ur during the Third Dynasty and early Old Babylonian periods: the AH Site. Roman numerals designate house numbers; arabic numerals have been assigned to each room. (Redrawn after Woolley 1974; Woolley and Mallowan 1976.)

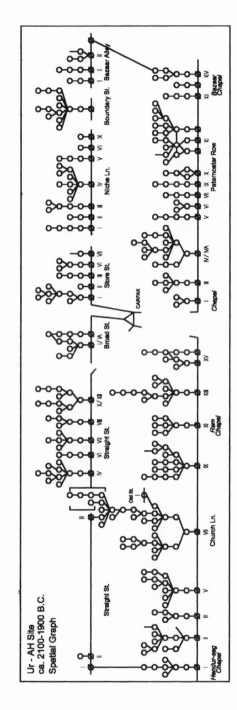

FIGURE 38. A spatial graph of the AH Site at Ur. (Source: author.)

to have been formed by linking a radial set of rooms (those on the left) with a linear block (those on the right). The most extensive example of this comes from VII Church Lane, which appears to have been assembled from three previously independent radial blocks that were conjoined to create a large nested circulation pattern with outlets on three different streets (Church Lane, Old Street, and Straight Street). Alternatively, the rooms at IVA Paternoster Row appear to have been subdivided from those of IV Paternoster Row when the communicating doorways were walled up (Woolley and Mallowan 1976: 147).

The general picture that emerges of the Ur residential areas, like that of the site as a whole, is one of rather intense fragmentation. Few of the routes through the precinct extend more than 30 or 40 meters before they confront a jog or a twist in the road. Indeed, even within the residences themselves some tension exists between the highly asymmetric use of extended room lines (linear pattern) and more symmetric distributions of residential space (radial pattern). As a result, there were significant fractures in the residential spaces of Ur both in the axes of transit and in the internal configurations of residential space. What is most telling about this configuration of urban neighborhoods are both the overall sense of experiential fragmentation and the points for easy surveillance and control of circulation that can be found within the interior space of the residence (the central court, the symmetric entryway) and within the axial network of the precinct (such as Carfax). Such supervision is, of course, a central component of authority but is particularly telling in this case because it provides a formal means of articulating the grassroots spaces of local neighborhoods with the domain of the household.

The fractured nature of urban experience in early southern Mesopotamia appears to be the subject of a rather humorous tale about a physician from the city of Isin who travels to Nippur, a center of Sumerian learning, to collect a fee:

"Where should I go in Nippur your city?"

"When you come to Nippur my [city], you should enter by the Grand Gate and leave a street, a boulevard, a square, [Til]lazida Street, and the ways of Nusku and Nininema to your left. You should ask [Nin-lugal]-apsu, daughter of Ki'agga-Enbilulu, [daughter-in-law] of Ninshu(?)-ana-Ea-takla, a gardening woman of the garden Henun-Enlil, sitting on the ground of Tillazida selling produce, and she will show you." (Foster 1996: 819)[39]

39. My thanks to Martha Roth for pointing this text out to me.

Although the main joke of the text centers on the physician's poor command of Sumerian, the directions given to him to find the one who will pay the fee are amusing in their superposition of a complex spatial map with an equally complex social map of the kin relations of the person he is to ask for directions.

The preceding discussion has sought to contrast the imagination and perception of early southern Mesopotamia urban landscapes with their experience. Where representations and built aesthetics portray cities as coherent, integrated, and singular places, an analysis of the physical space indicates a highly fragmented experiential landscape. However, what remains unclear are the implications of this multidimensional view of landscape for our understanding of the constitution of political authority in urban settings.

Authority and the Landscapes of Regimes

Given the evidence that Mesopotamian cities were profoundly political productions, it is rather counterintuitive that the dominant accounts of urban sites in southern Mesopotamia describe them as *villes spontanée* growing over time in ways rooted primarily in the strictures of the land and the daily life of the citizenry.[40] The resultant form is thus described as irregular, random, nongeometric, and "organic." As Van de Mieroop has argued, "The majority of [Mesopotamian] cities developed *naturally* over time from villages to urban centers, and showed no advanced planning in their layout. They were the result of urban growth over the centuries" (1997: 83, emphasis added). The distinction drawn between the *ville créée* and *ville spontanée* purports to describe a fundamental difference in spatial practices between cities with highly centralized apparatus for decision making about spatial development and cities where such decisions are made within more basal social groups, such as the family.

There are, however, a number of problems with this description generally and its application to southern Mesopotamian urban landscapes in particular. The most obvious objection is of course that no system of planning, whether accomplished by a town council or by a family or by a single individual, is more natural than another. In addition, the "organic" de-

40. Ferdinando Castagnoli (1971) defines the distinction between planned and organic cities this way: "The irregular city is the result of development left entirely to individuals who actually live on the land. If a governing body divides the land and disposes of it before it is handed over to the users, a uniformly patterned city will emerge."

scription of irregular cities often mistakes cultural variation in aesthetics for decentralization of urban planning. In searching for incidences of early planning in southern Mesopotamia, Van de Mieroop attributes only rectilinear public works, such as the rectangular walls of Babylon, to central planning. The attribution of the curvilinear to the unplanned reflects a distinctly modern and Western spatial aesthetic. For example, despite long providing the most resplendent example of an unplanned city, the gothic curves and sinuous streets of Siena, Italy, were codified and protected by the city council as early as A.D. 1346 (Kostoff 1991: 70). Thus the organic/planned distinction may be said to relate less directly to divergent spatial practices than to the dominant aesthetics of the observing subject.

Furthermore, the naturalistic overtones of the "organic" description tends to misplace the search for interpretations of form. In searching for an explanation for the dense packing of urban residential districts, Van de Mieroop cites the climatic advantages of having fewer walls exposed to the intense sun. Such environmental arguments are generally exceedingly flimsy because one needs only to point to, for example, the broad spacing of residences at the comparatively hot site of el-Amarna in Egypt or the densely packed houses of temperate medieval London to cast them in serious jeopardy. Furthermore, to attribute the aggregative character of buildings to climate does little to describe why they aggregate in the way that they do.

This is not to suggest that there is no difference between cities with stringent codified restrictions on the construction of new spaces and those that leave such decisions in the hands of variably empowered subjects or organizations. Anyone who has been to both Portland, Oregon, and Phoenix, Arizona, knows the difference rather intuitively between a city whose growth has been shaped by direct governmental regulation and one where planning for growth was deliberately ceded to real estate developers (who, not coincidentally, have long composed a significant portion of Arizona's governing institutions). But it is important to note that, in both cases, a political decision has been made: in the former to take direct control over a certain set of practical decisions, and in the latter to eschew such a move. Both, then, are in a very real sense planned; it is simply that their plans differ in the articulation of political practice with the experience of urban landscapes. The opposition is thus not between the planned and the organic but between various competing plans and their vision of the proper role of political authorities in landscape production.

Having established a place for politics within the production of southern Mesopotamian urban landscapes, such as Ur, the problem that remains for this discussion is where can we locate the primary political sources of

spatial production. The dominant locus of spatial production in southern Mesopotamia by the Ur III period was the king. The authority of the
southern Mesopotamian sovereign developed over the course of the third
millennium B.C. from rather poorly understood origins. Thorkild Jacobsen, one of the foremost epigraphers of the twentieth century, suggested that the king arose as a secular, primarily military, leader beside
the two preexisting loci of authority, the temple and the city assemblies
(one composed of city elders and the other convened as a general assembly
of citizens).[41] By the close of the third millennium B.C., the institution
of kingship had assumed the mantle of sovereign authority both within
the city and across the polity. This entailed a number of obligations, all
of which may be understood as shaping the landscape of the city and the
polity. The year-names given during Ur-Namma's reign provide a succinct encapsulation of both the critical obligations of kingship that Ur-
Namma hoped to highlight and the profound spatiality of these accomplishments (compiled from Frayne 1997: 10–18):

Year 1. Accession

Year 2. Installs daughter as the *en* priestess at Ur

Year 3. Raid on Lagash

Year 4. Expels the Gutians

Year 5a. Constructs the walls of Ur

Year 5b. Temple construction: Eridu, Ku'ar, and Ur

Year 6. Visit to Nippur

Year 7. Construction of the temenos and the Nanna complex

Year 8a. "Put the road in order"

Year 8b. Restores trade with Magan

Year 8c. Promulgation of the law code

Year 8d. Incorporation of new territory

Year 9. Installs *en* of Inanna at Uruk

Year 10. Chariot of Ninlil fashioned

Year 11. Builds Enlil temple at Nippur

Year 12. Digs Iturungal Canal

41. On the emergence of kingship in Mesopotamia, see Diakonoff 1969; Gelb 1969;
Jacobsen 1970; Postgate 1992; Steinkeller 1987.

The king's role as military leader placed him at the center of the geopolitical constitution of the regional order. As the guarantor of the fertility of the land, the king was responsible for digging and maintaining many of the large canals that irrigated the land and thus shaped the configuration of the countryside. But perhaps the king's most profound obligation was to safeguard the city and to propitiate the deities, responsibilities that he undertook as the architect of the urban landscape through extensive building programs focused on the monumental fabric of the city, most importantly the temples and city infrastructure.

The king's responsibility for city infrastructure is most poetically expressed by Gilgamesh's construction of the walls of Uruk. The role of the king as urban planner was a profoundly sacred role. The king's intercession was not only as provider of resources and labor but also, more directly, as the official charged with siting the construction on divinely established foundations. Thus the city's spatial infrastructure was established by the king but legitimated in reference to the gods. Royal works were critical in creating the urban landscape as a physical environment, establishing the broad parameters of circulation that divided the city and the built elements that evoked a sense of urban coherence (including the walls and gates that spatially defined the city and the pivotal institutional loci of the palace and major temples). As architect of the polity's canal system, the king would also have been responsible for the maintenance of the major infrastructure of the kārum and for the major city canals. It seems that the major thoroughfares of the city were often of royal design, though the best example of this comes from the much later processional way of Babylon (dated to the first millennium B.C.).

In mapping out the elements of the physical city that were produced by kings, it is clear that the urban landscapes of southern Mesopotamia were quite profoundly shaped by royal authority. However, the king was not the only locus of spatial production within the city. As Norman Yoffee (2000) has noted, sites of grassroots authority within early Mesopotamian cities such as the assemblies, the mayor *(rabiānum,)* the kārum, and in some cases the city elders *(puhrum,)* provided critical loci for meditating social conflicts, particularly those arising out of disputes over property and inheritance. Bills of sale for property recovered in the AH site testify to juridical supervision of contracts, and the law code of Hammurabi stipulates a wide range of issues in the structuring of city space that fell under the authority of the ruler to command and adjudicate. Several of the "laws" describe regulations bearing on relations between husband and wife, father and heirs, and buyer and seller. These allow for the intervention of political authorities in the division and alienation of property that holds profound consequences for the production and reproduction of residential space (see Roth 1995a: esp. laws nos. 148, 150, 165, 166, 168, 169, 170, 178, 179, 180).

The extra-royal administration of the city appears to have been organized across the city as a whole through the city assemblies and within individual wards, quarters, or neighborhoods—local assemblies of residents charged with regulating certain affairs "from sanitation to security" (Oppenheim 1969: 9). City assemblies are a rather poorly understood political institution in early southern Mesopotamia. It appears that assemblies existed both for the city as a whole and for each of the city quarters. In general, city quarter or ward assemblies seem to have served largely juridical functions, as courts for adjudicating disputes. The power of the assemblies to produce elements of the urban landscape was far less direct than that of the king yet potentially more intrusive into the everyday life of subjects. As Oppenheim (ibid.) has noted, contracts between private parties were witnessed by city and royal officials. And the law codes of both Shulgi and Hammurabi described the authority of assemblies, particularly those of the wards, to resolve inheritance disputes. This authority provided these more grassroots elements of urban political regimes with considerable power to shape both neighborhood and domestic space.

Ethnoarchaeological studies by Carol Kramer (1982) and Lee Horne (1994) have suggested that the inheritance of houses and purchase of new spaces play a critical role in shaping urban form in both the modern and the ancient Near East. We see some evidence for this in the case of the AH site at Ur, where walls built between previously unified rooms suggest a division of the household and the house. Similarly, purchase of additional spaces, such as for storage, need not be in adjacent spaces but might be elsewhere in the neighborhood. The result is a number of very small rooms that are insufficient for an independent household but might have been sufficient as supplemental storage space. But the rights of inheritance and alienation were subject to the regulation of the sovereign and the courts.

Another critical locus of political power within early Mesopotamian cities was the mayor. The role of the mayor, as described by Rivkah Harris (1975) in her study of Sippar, was to act as mediator between the palace and subjects within a ward. Most notable among these responsibilities was to act as a middleman in the flow of revenue between subjects and king. Texts uncovered during excavations in a mayor's house at the Babylonian outpost at Haradum indicate that this mediating position was a source of considerable power and wealth, even though it was not unchecked: "Concerning the silver, which Habasanu during his tenure as mayor had made the town pay, the entire town assembled and spoke in these terms to Habasanu: 'of the silver which you made us pay, a great amount has stayed in your house, as well as the sheep which we gave on top as voluntary gifts.'" (Van de Mieroop 1997: 130). Thus, the power of the mayor to shape the physical environment of the city was predominantly located in his position as a mediator in the flow of capital that allowed for royal construction programs. This transfer of resources from subjects to elites marked another critical point of control for spatial production.

Even from this necessarily cursory sociology of urban political authority, it is clear that, though the production of the urban landscape was most profoundly sited in the king (as both direct sponsor of building programs and as juridical authority of last resort in cases of property exchange and inheritance), he was not its sole producer—a stark contrast to the vision of the city forwarded in royal media. At every point within the city, variously sited elements of the ruling regime possessed the authority to intervene in or directly organize the production of spatial form and hence the experience of the urban landscape. The king played a role in the creation of the city as did the mayors, judges, and assemblies, both local and city-wide. Indeed, the subjects of the city are themselves profoundly implicated in the creation of the city. Even though the royal apparatus dominated the production of the spatial imaginary through its privileged controls over monumental media, the production of urban spatial experience cannot be located simply in an elitist account of urban politics centered on the king. Rather, the political sources must be broadened to incorporate regimes—coalitions of critically located authorities sited in relation to both intra-elite ties (such as king to temple priesthood) and links between the sovereign and more grassroots organizations (such as the assemblies; Mollenkopf 1992). Thus, as Socrates argued in Plato's *The Republic,* we would indeed be wrong to think of Mesopotamian urban landscapes as one city rather than many cities.

I have not argued in this chapter that there was no such thing as the Mesopotamian City, a place that endured both across the space of the alluvium and three millennia of history—indeed there was. However, it was an imagined place, produced initially within politically generated representations and subsequently within the modern academic reductionism of the ideal type. By reiterating this rather singular representation, contemporary scholarship has tended to obscure the operation of the political in the production of distinct landscapes and preclude the teasing out of the imagined and perceived from the experienced. The coherence attributed to the ancient Mesopotamian City was a political production, an image of the urban political landscape promoted by regimes, even as experiences of city life came to be constituted by a host of divisions and fragmentations that secured the reproduction of authority. But just as the coherence of the city was largely illusory, regimes themselves tended to be riven with fractures. In the next chapter, I will turn our attention to the role that landscapes play in constituting authority within relationships among political institutions.

CHAPTER 6

Institutions

Since time immemorial. . . . The oxen of the gods ploughed the
garlic plot of the ruler, and the best fields of the gods became the
garlic and cucumber plots of the ruler. Teams of asses and spirited
oxen were yoked for the temple administrators but the grain of
the temple administrators was distributed by the personnel of the
ruler. . . . The administrators felled trees in the orchards of the poor
and bundled off the fruit. . . . The bureaucracy was operating from
the boundary of Ningirsu to the Sea. . . . These were the conventions
of former times.

"Reforms of Uruinimgina [Urukagina]"
of Lagash, twenty-fourth century B.C.
(from J. S. Cooper, *Presargonic Inscription*)

The tumult during the 1990s over the rebuilding of the Federal Chancellery in Berlin has attracted much attention because of the intense political symbolism at play both in the transfer of the seat of German government back to Berlin and in the form and aesthetics of the building itself
(fig. 39a). Two elements of the design have come in for the most intense
scrutiny. First, its sheer size has inspired some unease in pundits, European politicians, and architectural critics insofar as it appears to reinscribe
Albert Speer's monumental vision for the political center of an imperial
Nazi Germany, or at the very least provide a ridiculously grandiose statement on the power of Helmut Kohl's regime (the original commissioner
of the project) to effect a united Germany out of the ashes of the Cold
War (Cohen 2001). In contrast, the central glass dome of the renovated

Reichstag has been praised by many of the same writers for formalizing a transparency to the operation of government in keeping with liberal democratic ideology (fig. 39b). Thus, much of the tension generated by Axel Schultes's design for the Chancellery and Norman Foster's vision of the Reichstag centers on their ambivalent deployment of aesthetic features from Germany's fascist past as well as its democratic present.

The debate over the design of Berlin's new federal architecture has been waged almost solely over the expressivity of its architectural aesthetics.[1] The implicit assumption organizing much of the debate seems to be that buildings in and of themselves can be fascist or democratic, or at least emote fascist or democratic values. This form of architectural absolutism is predicated on the contention that the aesthetics of built political institutions, places that might in large measure come under Charles Goodsell's (1988) term "civic space," carry with them specific forms of governmental authority and that this authority is both singular and coherent. But the expressivity of the institutional architecture of government is most assuredly ideological in that it obscures as much about the operation of politics as it reveals. Underneath the formal aesthetics of the Chancellery lies a set of complex relationships among a host of governmental institutions—rival parties, rival bureaucracies, rival seats of power—that are actively producing a new German capital in reference to an old set of places. What is most intriguing about the new Chancellery is the manner in which it is situated within a broadly reconfigured political landscape of German governmental institutions that not only is highly evocative but also promises to generate new physical relationships among major seats of power and a transformed imagination of the proper institutional order of political authority.

Although political institutions emerged as critical structural elements of the State within early modernist political theory (see Durkheim 1986; Gramsci 1971; Weber 1968), they were largely eclipsed during much of the later twentieth century, in part by contextualist sociologies that highlighted the impact of sociocultural divisions (class, ideology, ethnicity, language, religion, gender) on a broadly configured governmental apparatus (see Dahrendorf 1959; Easton 1968; Lipset 1959; MacKinnon 1989; Parsons 1951) and by behavioralist perspectives that tended to reduce politics to aggregates of individual or group decisions (such as "the Market"; see Niskanen 1971; Stigler 1952). A recent revival of institutionalist perspectives

1. For an excellent overview of the architectural history of Berlin and its impact on current debates, see Wise 1998.

FIGURE 39. The new architecture of the German federal regime, Berlin. **a.** Axel Schultes's new Chancellery. **b.** Norman Foster's redesigned Reichstag. (Photos courtesy of Lisa Weeber and Martina Frank.)

in political science has brought the organizational apparatus of governance back into considerations of politics as both privileged sites of political action (or inaction) and arenas for contending social forces (March and Olsen 1989: 17–19, 1995; Peters 1999; Soltan 1996; see also Evans, Rueschemeyer, and Skocpol 1985; Green 1996). For James March and Johan Olsen, early advocates of the so-called "new institutionalism," the institution is less a formal structure than a set of enduring procedures, routines, and values that establish the frameworks within which social and political relationships proceed. Highly critical of movements in political thought that have foregrounded the (rational) individual at the expense of an adequate theorization of the collective routines at the heart of political formation and reproduction, March and Olsen argue for an account of politics centered in relationships among collectivities. Institutions—collectivities bound together by shared histories and interests that shape ingrained values and routines—recursively shape their members and, over time, can provide the foundations for governmental stability (or ossification) and transformation.

March and Olsen's efforts to dissolve the longstanding dispute in political science and sociology between the social and the governmental through a revived emphasis on an array of institutions as loci of collective action have been quite productive. However, their emphasis on institutions as primarily discursive fields, where a logic of "appropriateness" constrains decisions and actions, tends to ignore the fundamental difference between institutions and less regular social associations and organizations. That is, political institutions are profoundly sited in place within an architectural landscape that draws together not only discourses on appropriate action but also physical demands on inter-institutional ties and imaginings of the governmental apparatus as a whole. In the seventeenth century, Christopher Wren called attention to the power of public architecture: "Architecture has its political use, publick buildings being the ornament of a country. . . . It establishes a nation, draws people and commerce; makes the people love their native country" (quoted in Tinniswood 1998: 7). But Wren still underestimated the importance of institutional architecture to the operation of politics and the constitution of authority. The sense of power evoked by the panels depicting Assyrian conquests in Sennacherib's palace at Nineveh (Russell 1991) or by the exhausting trek from the entrance of Speer's Chancellery to Hitler's office certainly bolsters the regime as a whole but more narrowly establishes the authority of certain offices and institutions (such as kingship and an executive bureaucracy) within that regime. Hence, when describing his vision for Speer's Chancellery, Hitler emphasized his own personal authority, demanding that when entering the building "one should have the feeling that one is visiting the master of the world" (Tinniswood 1998: 155). This political relationship among institutions was also foremost in Pierre Charles L'Enfant's plan for Washington, D.C. At the center of his Baroque design, a right triangle cut by diagonal streets links the presidential palace (now the White House) to the Washington monument (originally planned as an equestrian statue of the first president) and the Capitol building. The Capitol building and the presidential palace were both sited on elevated terrain, in the grand manner, to emphasize their collective authority, but the oblique distance between the two buildings formally inscribed their rivalry as intertwined loci of institutional power. To overlook the places of these institutions in favor of purely discursive routines is to miss something critical about the operation and imagination of their authority. Moreover, to focus purely on the aesthetic expressivity of these buildings is to ignore the real ways in which these institutions are positioned within the political landscape.

The reforms enacted by Uruinimgina, the last king of the first dynasty of the city of Lagash in southern Mesopotamia, provide a glimpse into the institutional relationships that frame not only the evocative expressivity of architectural aesthetics but also the experience and imagination of institutional ties within an early complex polity.[2] The text of the reforms is divided into four major parts, all of which are written in the third person rather than in the more boastful first-person style that came to dominate political rhetoric after Sargon. In the first section, the author lists several of the king's building projects, telling of his efforts to support temples and extend canals. In the second section of the text, excerpted in the opening epigraph to this chapter, the author provides a list of injustices visited on temples by rulers. Following the inventory of abuses heaped on temples (or, more precisely, temple administrators) by kings from "time immemorial," the texts detail a series of heroic reforms enacted by Uruinimgina that formally established limits on the ability of the king to intrude on the facilities and resources of the temples:

> When Ningirsu, warrior of Enlil, granted the kingship of Lagash to Uruinimgina . . . he replaced the customs of former times . . . he removed the silo supervisor from (control over) the grain taxes of the *guda*-priests . . . and he removed the bureaucrat (responsible) for the delivery of duties by the temple administrators to the palace. He installed [the god] Ningirsu as proprietor over

2. The reforms of Uruinimgina (a.k.a. Urukagina) are known to us from six inscriptions that provide three essentially identical versions of the same text (Foster 1981: 230; see also Maekawa 1973–74).

the ruler's estate and the ruler's fields; he installed [the goddess] Bau as proprietor of the estate of the women's organization . . . ; and he installed Shul-
shagana [the child of Ningirsu and Bau] as proprietor of the children's estate.
From the boundary of Ningirsu to the sea, the bureaucracy ceased operations. . . .
He cleared and cancelled obligations for those indentured families, citizens of
Lagash living as debtors because of grain taxes, barley payments, theft or murder. Uruinimgina solemnly promised Ningirsu that he would never subjugate
the waif and the widow to the powerful. (Cooper 1986: 71–73)

The fourth and final section of the text returns to the architectural theme
that opened the composition, describing a number of canals built by the
king in the countryside.

The image of a predatory bureaucracy dismantled holds a rather transparent allure for modern readers, and the early interpreters of the reform
texts came to understand them in terms highly reminiscent of early-
twentieth-century crises of authority. Anton Deimel (1931) and Anna
Schneider (1920) interpreted the reforms of Uruinimgina as a reaction
against the increasing power of secular political authority, vested in the king,
that had expanded in Lagash during the preceding reigns of the first dynasty. They read in the texts a tale of restoration, a reassertion by the temples
of their traditional rights and their primacy within the Sumerian institutional alignment. Benjamin Foster (1981: 235) has cast serious doubt on this
interpretation, pointing out that the indignities rulers visited on temples
dated back to "time immemorial" and were not a recent assault on traditional institutional prerogatives. Indeed, it seems clear that Uruinimgina
did not describe his reforms as a reactionary turn back to a more pious past
but as a radical transformation, an attack on long-practiced injustices undertaken at the behest of Ningirsu, the patron deity of the city. Whether radical or reactionary, the reforms of Uruinimgina do provide a scintillating
account of a profound institutional rivalry within the mid-third millennium
B.C. politics of Lagash, a window onto a moment of intra-regime contestation between religious and regal sites of authority. It is important to
note that Uruinimgina bracketed his account of the reforms with descriptions of architectural accomplishments—expansions of temples, extensions
of canals—that inscribed elements of the reforms in the political landscape.

Archaeologists have long utilized architectural forms as a shorthand
for major sociopolitical institutions. Hence temple, palace, and market-
place regularly stand in for religious, royal, and economic institutions.
But the link between architecture and institution has traditionally been
purely synecdochic, a tropic convention rather than a well-theorized vision of political architectonics. In this chapter, I examine the production
of institutional relationships within political landscapes, focusing again

on the empire of Urartu and in particular on the apparatus of governance that it constructed on the Ararat plain during the eighth and seventh centuries B.C. The goal of this discussion is not to reformalize institutions as big stately buildings. Rather, the following analyses are intended to examine how the relationships and routines that March and Olsen describe as constitutive of political institutions are profoundly negotiated within landscapes—in physical architectonics that organize relationships among organizational loci, in varying imagined visions of the landscape that contest the terms in which political legitimacy is sought, and in the perception of institutional architectures that efface these divisions.

Institutions and Urartian Political Landscapes

As Paul Zimansky has cogently noted, one frustrating peculiarity of the study of Urartian institutions is that the two primary sources of data—epigraphic and archaeological—tend to "illuminate different phases in the history of the kingdom" (1998: 70). Although the long royal display inscriptions that record military campaigns, deities, and aspects of the political economy were largely composed during Urartu's era of high imperialism during the eighth century B.C., the majority of the archaeological record comes from sites established by the energetic King Rusa II of Urartu's seventh century B.C. period of reconstruction that followed the disruptions sparked by Sargon II's eighth campaign (see chapter 4). The kingdom's provinces in the Ararat plain present one important exception to this generalization. Major sites have been excavated there that date to both the imperial and the reconstruction periods. Erebuni and Argishtihinili, sites founded by Argishti I immediately following his conquest of the region (by 780 B.C.), have both hosted extensive excavations that have revealed a great deal of the architectural plan of the two sites. Teishebai URU (a.k.a. Karmir-Blur), in contrast, was established by Rusa II during the seventh century B.C., apparently as a regional center to replace Erebuni.

The basic institutional outlines of the Urartian polity are rather poorly understood at present, largely because of the royal monopoly on display inscriptions that leads us to view Urartu through the monochromatic lens of the king. However, the sheer scale of Urartian political centers and the resources that they housed suggests an extensive governmental apparatus. The elements within this apparatus and the relationships defined between them within the developing architecture and imagery of the Urartian polity are the focus of the following discussions.

PERCEPTION

The perception of regimes as coherent and unified across the political land-scape (discussed in chapter 5) in large measure creates the central problem of this chapter. If regimes build a sense of their authority as coterminous with key places—the city, the nation, the polity—then how can we tease out the operation of distinct institutions within political landscapes? Urartian political centers were produced out of a very different environmental aesthetic than early cities in southern Mesopotamia. Where Mesopotamian political centers were situated within sprawling urban landscapes that hosted a diverse array of activities, Urartian political centers were much more limited in the social practices that they enclosed. The range of practices that took place within Urartian fortresses was largely confined to those immediately bearing on the operation of the political apparatus. Thus, though manufacturing the perception of coherence out of heterogeneous places was a serious problem for the rulers of Ur, Urartian centers were predicated much more forcefully on establishing the essential architectural unity of the governing regime. In this respect, Urartian fortresses closely followed the traditional architectonics of political centers established in the highlands of eastern Anatolia and southern Transcaucasia beginning in the mid-second millennium B.C. (Smith and Thompson forthcoming).

Like these early political centers, the ashlar masonry walls of Urartian fortresses circumscribed only the immediate apparatus of governance, evoking a very clear sense of the internal integrity and coherence of the ruling regime. Furthermore, the regularities in siting and design of Urartian fortresses established a clear marking of place in reference to an over-arching supra-regional polity, rather than to traditional expressions of the local. Variation certainly existed within the masonry styles of different regions (fig. 40), setting apart, for example, the less well dressed stone masonry of Argishtihinili from the regular fitted blocks of Van Kale. Furthermore, variability in masonry styles between eighth century B.C. fortresses, such as Argishtihinili and Erebuni, as compared to seventh century B.C. sites such as Teishebai URU and Bastam suggests an increasing concern to standardize the environmental aesthetics of Urartian political centers. Whereas earlier fortresses around the kingdom boasted an intriguing range in their workmanship, materials, and sensuality of form, the later constructions of Rusa II are more homogeneous in their architectural expressivity. This trend is in direct contrast with contemporary transformations in the experience of Urartian fortresses. As the Urartian regime tried to promote a more standardized evocative aesthetics of pub-

FIGURE 40. Comparative views of Urartian masonry. **a.** Horom (north citadel, fortification wall B). **b.** Erebuni (west fortification wall, all but lower two courses are reconstructed). **c.** Ayanis (gateway area). **d.** Argishtihinili (southern fortification wall). **e.** Lower Anzaf (west fortification wall). **f.** Van Kale (only lower ashlar masonry courses are Urartian constructions). **g.** Çavuştepe (East Main Gate). (All photographs by author.)

lic architecture, the physical organization of the fortresses became significantly less coherent, suggesting shifting practical relations among Urartu's political institutions.

EXPERIENCE

The organization of space within imperial period Urartian fortresses on the Ararat plain is relatively well documented thanks to extensive excavations at the sites of Erebuni and Argishtihinili. It is quite apparent that Urartian fortresses were built environments that assembled spaces in very particular ways to define sets of spatial relations between functionally discrete components of a complex political apparatus. These components emerge most clearly when we compare the local development of the physical institutional landscape over time, from the eighth century B.C. centers of Erebuni and Argishtihinili to the massive fortified complex of Teishebai URU built by Rusa II. In moving beyond the exclusively perceptual terms in which institutional architecture is traditionally discussed, I must examine the experiential dimension of these three fortresses in some detail. It is only in these details that we can track the slight, yet significant, shifts in the formal organization of Urartian institutions.

Argishtihinili. Located 5 kilometers north of the Araxes river, the large Urartian center of Argishtihinili occupied an extended ridge, known as the Hills of David, and an adjacent cone-shaped basalt outcrop 2.5 kilometers to the east, known as the Hill of Armavir. Fortress walls at the site varied in thickness from 2 to 5 meters and were constructed of basalt blocks with a mud-brick superstructure. As described above, the stone masonry was semi-ashlar in style, with dressed stones set into defined courses but without the tightly fitted joints known from contemporary Urartian constructions in the Lake Van area.

The western fortress at Argishtihinili, the most thoroughly explored of the two complexes, was built atop the largest of the five rises that constitute the Hills of David (fig. 41). The main gate into the fortress lay on the northern slope and was approached via a ramp of packed earth (Kafadarian 1984: 134). The internal organization of space, though not entirely known because of the differential preservation of walls across the site, is nonetheless substantially clear in the large exposures made by the excavators. A spatial graph of the western fortress at Argishtihinili (fig. 42) affords an excellent view as to how the spaces in the fortress were assembled. It is immediately clear from this graph that the fortress was com-

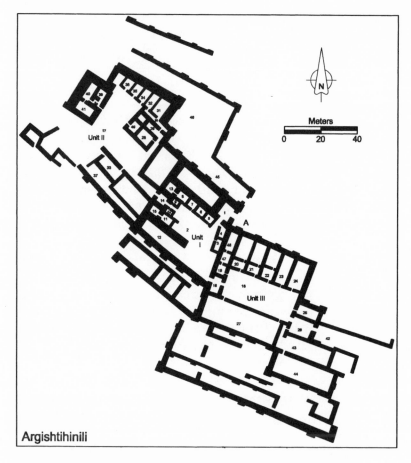

FIGURE 41. Architectural plan of Argishtihinili West. (Source: Institute of Archaeology and Ethnography, Republic of Armenia.)

posed of three largely independent units, organized around sets of small rooms radiating from central courtyards. These units were linked to each other via a single pathway, suggesting that movement between units was highly regulated while circulation within the units was relatively open.

The architectonic relationships within the fortress can be quantified from the spatial graphs along two axes of particular interest: symmetry and distribution.[3] Spaces are defined as symmetric when pathways link all spaces;

3. The procedures of spatial analysis described here were originally defined by Bill Hillier and Julienne Hanson (1984). These techniques have since been revised by T. A. Markus (1993), who divorced them from Hillier and Hanson's problematic generative theory of

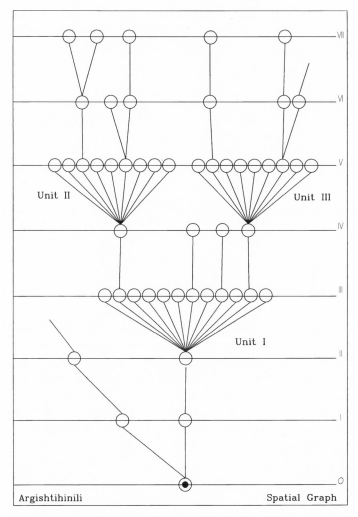

FIGURE 42. A spatial graph of Argishtihinili West. (Source: author.)

asymmetry, in contrast, implies spaces that are accessible only through others. Spaces are distributed when they are linked by more than one nonintersecting pathway; nondistributed spaces, conversely, have few pathways, creating branching structures with numerous dead-ends. Symmetry can

social space. Underlying these techniques are two key assumptions: that space is a continuous, structured aspect of daily practice, and that differential access to spaces, their "permeability," embeds differential sociopolitical power in built space.

be quantified by measuring the depth of a space from all other spaces in the system. Real relative asymmetry (RRA) values compare actual depth with how deep a built environment could theoretically be given the total number of spaces.[4] RRA values are both above and below 1.0. Low RRA values (less than 1.0) indicate a relatively integrating built environment. The measure of distribution (or control) quantifies the number of neighbors for each space relative to the neighbors of each adjacent space. Each space gives $1/n$ to its neighbors, where n equals the number of adjacent spaces. The values received by each space from its neighbors are then summed to give the control value for that space. Spaces with control values greater than 1.0 indicate a nondistributed space, one in which control is potentially strong. Although no quantitative measure can completely describe the spaces that they purport to represent, nor are they appropriate to use in all cases, they give us a method for defining transformations in architectonics over time that can yield insights into spatial production.

The RRA value for Argishtihinili fortress as a whole is a moderate 1.0483, indicating that the architectonics of the fortress were moderately symmetrical suggestive of a relatively integrating built environment. Control values for the fortress suggest that rooms 2, 17, and 18 regulated access between each unit (see fig. 41). Thus, though the spatial organization of the fortress overall was moderately integrating, control values indicate potentially rigorous oversight of movement between architectonically discrete units.

Excavations at Argishtihinili West have shed considerable light on the primary uses of space within each unit. Unit I (rooms 2–15) was used as living space for "subordinate administrators"; according to the excavator, the central courtyard and adjacent units quartered low-level bureaucratic functionaries (in contrast to the residences of higher government authorities found outside the fortress; Martirosian 1974: 90).

The foci of Unit II (rooms 17, 29–41, 49) were three buildings described

4. RRA values are calculated by first defining the mean depth (MD) for the system from a given point by assigning depth values to every other space in the system depending on the number of steps away from the original point. Thus, all spaces adjacent to the original point will have a depth value of 1, those one step away will have a value of 2, and so on. The mean depth from that point can then be calculated by summing the depth values and dividing by the number of spaces in the system (k), minus 1 (the original space). With mean depth calculated, the relative asymmetry (RA) value for a space can be worked out using the following formula: $RA = 2(MD - 1)/k - 2$. RA values are between 0 and 1. For these values to be comparable with other built environments of different sizes, each RA value must be divided by a constant tabulated by Hillier and Hanson (1984: 112) to produce RRA values. See Ferguson 1996 and Markus 1993 for more detailed discussions of procedures for calculating asymmetry and control.

by the excavator as temples (ibid., 86–87). Temples 1 and 2 (rooms 28, 29, 49) are rather nondescript constructions, identified as religious spaces largely based on the prevalence of ritual objects such as specialized vessels, wall lamps, and iron wall nails among the recovered materials. The religious nature of Temple 3 (rooms 38–41) was inferred from both the extant assemblage and architectural parallels with Mesopotamian (not other Urartian) temples (ibid., 86; see also Forbes 1983). This solidly built complex of four rooms, though incompletely excavated, was constructed of finely dressed basalt blocks set into walls from 2.5 meters to 5.0 meters thick. The doorway into the structure was constructed of two massive upright basalt blocks decorated with smooth plaster and a basalt lintel. The remaining mud-brick superstructure carried traces of painted plaster.

Unit III (rooms 16, 18–27, 42–48), the eastern wing of the fortress, was also the largest. Although incompletely excavated, those rooms that were investigated yielded evidence of extensive storage facilities. Room 43 contained rows of large pithoi set into the floor, similar to those documented in store rooms at other Urartian sites such as Bastam (Kleiss 1977: 30–32). A similar arrangement was found in the excavated portion of room 27, suggesting that all of the rooms in Unit III were involved in storage of imperial revenues.

The three architectonic units of the fortress provide an outline of the Urartian political institutions in the Ararat plain, encompassing religious/temple and royal/bureaucratic spheres, with spatially distinct facilities for storing and redistributing goods. These same components are in evidence at the contemporary fortress of Erebuni.

Erebuni. The first Urartian fortress built north of the Araxes river, Erebuni (modern Arin-berd) overlooks a now densely settled area of the modern city of Yerevan.[5] Although Argishtihinili is often described as Urartu's regional economic center, Erebuni (46.7 kilometers to the northeast), has been contrasted as the Urartian regional political center during the eighth century B.C. (fig. 43). Partial reconstruction of the site during both the Achaemenid and the modern eras has introduced some problems into our understanding of the organization of the site as it was originally built.[6] Although the overall plan of the fortress is not thought to have been dra-

5. Details regarding the excavations at Erebuni are reported in Oganesian 1961.

6. There is a great deal of uncertainty at present regarding how faithfully the fortifications at Erebuni were restored to their Urartian era plan. Recent small-scale investigations near the exterior of the fortress wall suggest that Urartian levels at the site may not have been fully exposed or explored to their full depth during the original excavations at the site (Ter-

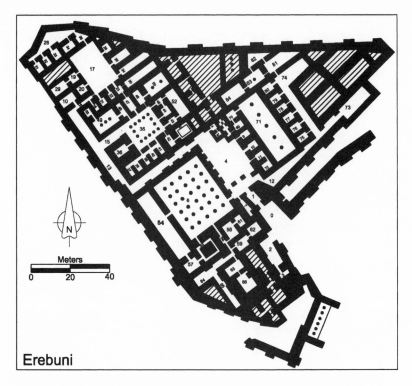

FIGURE 43. Architectural plan of Erebuni. (Source: Commission for the Preservation of Historical Monuments, Republic of Armenia.)

matically altered, some uncertainty lingers over the outline of the fortress walls and the dating of the columned hall at the center of the complex (room 6). Analysis of the organization of space at Erebuni is further hampered by the relative paucity of archaeological finds at the site, resulting from its abandonment in the seventh century B.C.

Although it occupied approximately half the surface area of Argishtihinili, Erebuni was composed of almost twice as many rooms. Built as a single construction unit, Erebuni gives the impression of a compact, tightly planned settlement. The spatial graph of the fortress (fig. 44) indicates that, though considerably deeper than Argishtihinili, Erebuni was organized into similar sets of interlinked but self-contained units. The mean RRA value of Erebuni is 1.327, somewhat higher than Argishtihinili,

Martirosov personal communication 2000). However, the existing plan remains our best basis for describing the Urartian period organization of the fortress.

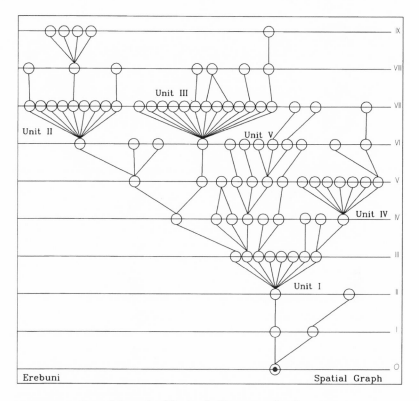

FIGURE 44. A spatial graph of Erebuni. (Source: author.)

reflecting a slightly greater tendency of Erebuni's spaces to isolate rather than integrate. Almost all of this increase in the mean RRA value for the fortress can be traced to the highly asymmetrical organization of Unit V. Perhaps the most striking aspect of the layout of Erebuni is the distribution of space. Movement through the fortress was tightly regulated by several access points that yielded high control values (rooms 4, 17, 35, 74).

The four branching complexes at Erebuni apparently were divided along functional lines. Unit I (rooms 4–11, 67, 72) served largely as a reception area, a feature conspicuously absent from Argishtihinili. The rooms of Unit II (rooms 14, 17, 19–26, 28–34) were incompletely excavated, and the uses of these spaces are not well understood. Although Boris Piotrovskii (1969: 70) suggests that they may have been storage facilities, analogy with Argishtihinili suggests that this unit could also have provided quarters for low-level government functionaries.

Unit III (rooms 15, 35–52) was a sacred precinct centered around the

"Susi" temple (room 36). The temple has been identified by two in situ dedicatory inscriptions found on either side of the entryway (Melikishvili 1971: nos. 396–97). The remainder of rooms surrounding the courtyard likely composed a residential complex for those associated with the temple, though the published excavation reports do not detail the contents of each room.

Unit IV (rooms 68–71, 73–84), described by Piotrovskii as a set of storage rooms, included a large central pillared room (71) that served as a wine cellar. The smaller rooms on either side of the wine cellar, Piotrovskii (1969) suggests, were granaries.

Although the branching complexes at Erebuni closely resemble those of Argishtihinili, the organization of Unit V (rooms 56, 58–66) departs significantly from this pattern of moderate asymmetry and strongly nondistributed space. Unit V was entered via room 6. Piotrovskii suggests that this room was originally a colonnade with two rows of columns that was later reconstructed ("perhaps in the Urartian period, perhaps later" [ibid., 70]). During reconstruction, the colonnade was extended and enclosed by a buttressed wall, producing a square hall with 30 columns that resembles apadanas known from later Achaemenid sites such as Persepolis, Susa, and Pasargadae (Tadgell 1998). This colonnade led into the royal residence, marked by the decorative paintings found on the rear wall (Oganesian 1961; Piotrovskii 1969; on the paintings at Erebuni, see Oganesian 1973). The royal residence is noticeable in its unique organization of residential space. As the spatial graph of the site illustrates, the rooms of Unit V are highly asymmetric and nondistributed, suggestive of both rigorous control and low integration. These impressions are confirmed by the elevated RRA and control values for the rooms in this complex.

Overall, the fortress of Erebuni displays a compartmentalization of space similar to that found at Argishtihinili with the addition of a complex of residential rooms. Although inscriptions emphasize that Erebuni was built as a base for further military excursions in the north, it was by no means simply a garrison town. In evidence at the site were the same political institutions—religious/temple, royal/bureaucratic, and redistributive/economic—also visible in the architectonics of Argishtihinili.

· · ·

Unfortunately, it is impossible at present to directly correlate the architectonic divisions visible at Argishtihinili and Erebuni with the textually known elements of Urartian administration delineated by Zimansky (primarily from seventh century B.C. sources; 1985: 77–94). However, two observations on the articulation of archaeological and historical records are worth mentioning. First, the epigraphic record is skewed toward royal activities and thus, as suggested by the architectonic analysis, describes only one of the institutions occupying Urartian fortresses. This would suggest that a full understanding of Urartian political organization cannot arise solely from textual sources. Second, the division between royal and provincial administration noted by Zimansky may be geographically variable (ibid., 89). That is, though inscriptions from the Assyrian border area describe provincial administrators distinct from the royal bureaucracy, on the Ararat plain there is little archaeological evidence for a division within imperial structure between a royal administration and a semi-redundant set of provincial administrators.

The experiential landscape produced by Urartian institutions in the Ararat plain during the imperial period facilitated not only further expansion but also the administration of the region by a centralized triumvirate of institutions: bureaucratic/royal, religious/temple, and distributive/economic. Unlike Assyrian or Persian imperial programs, the entire complex of Urartian institutions seems to have been part of a singular, highly integrated governmental package that followed conquest and occupation. The organization of space within the fortresses themselves suggests that religious, bureaucratic, and economic institutions, though maintaining a degree of structural independence indicated by the rigorous control exercised over movement between segments of the fortress, were well integrated into a singular political entity. There is nothing to suggest that any institution was significantly differentiated within the overall authority structure.

During the reconstruction period, though the cast of institutional characters does not seem to have changed significantly (suggesting continuity in Urartian political structure), the formal relations between them appear to have altered. Teishebai URU was built along different architectonic lines than either Erebuni or Argishtihinili.

Teishebai URU. With the exception of two small late additions, the fortress at Teishebai URU ("City of Teisheba")[7] was built in a single episode

---

7. Accounts of the excavations at Teishebai URU (formerly read "Teishebaini") can be found in several preliminary reports (e.g., Piotrovskii 1955) as well as in more synthetic accounts (e.g., Oganesian 1955; Piotrovskii 1959, 1969).

as one continuous building.[8] It consists of two discrete architectural units (fig. 45): the main fortress and an open courtyard to the west, circumscribed by an outer defensive wall. Entrance into the fortress was via two gateways defined by the intersection of the two units—the main gate to the south and the "postern" gate to the north (Piotrovskii 1969: 137). The main gateway, on the southern end of the courtyard, was flanked by two massive towers with a constricted entryway similar to the King's Gate at the Hittite capital, Hattusas (Kafadarian 1984: 64). The postern gate was defended only by a single tower on the northeastern corner of the fortress.

The fortification walls at Teishebai URU consisted of massive straight, rectilinear walls punctuated by large square towers and buttresses set at regular intervals. The fortifications were built using cut-stone boulder facades with a rubble core. The exterior fortress walls were constructed of a massive stone socle about 2.0 meters high (four courses of stones) and 3.5 meters thick capped by a mud-brick superstructure. The ground floor not only served as the platform for the second story but also contained numerous rooms. The interior walls of the fortress, in contrast to those of Erebuni, were built largely of mud-brick. The ground floor of the fortress was constructed on different levels, utilizing the uneven topography of the site to create a stepped arrangement. The rooms in the plan of the site represent only this ground floor; unfortunately, it is not possible given the available data to clearly define the organization of rooms in the second story.

Although the plan of a small area of the southern portion of the fortress remains undefined, 130 discrete rooms of the ground floor, as well as the doorways that connected them, have been mapped.[9] Because only the spatial organization of the ground floor has been preserved, the architectonics of Teishebai URU must be approximated from the extant data. However, the pathways that linked the rooms of the ground floor divide the fortress into three architectonically distinct units: one in the northwestern third of the fortress, a second in the center, and a third, only partially defined, in the south.

Figure 46 provides a partial spatial graph for the ground floor of the fortress. In this graph, access points (depth o) do not indicate the exterior of the fortress (as in the spatial graphs of Erebuni and Argishtihinili). In-

8. Although K. L. Oganesian (1955: 37) is not entirely clear as to which parts of the fortress he regards as extensions, one is certainly the small enclosure added to the western fortification wall.

9. The number of these rooms that were completely excavated is unclear from the site reports. Oganesian (1955: 40) reported that 65 rooms had been explored.

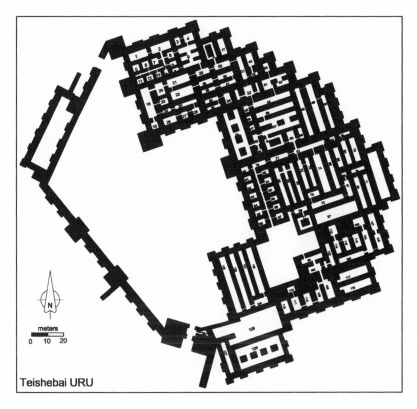

FIGURE 45. Architectural plan of Teishebai URU. (Source: Institute
of Archaeology and Ethnography, Republic of Armenia.)

stead, they refer to discrete locations in the upper floor through which the
lower floor would have been accessed (these access points are marked by
the letters A, B, and C in fig. 45). No doorway was found leading into the
ground floor from the courtyard, indicating that the only entrance to the
fortress proper was via an earthen ramp adjacent to the main gateway.[10]

   Although the lack of information regarding the layout of the upper
floors impedes understanding of the organization of the fortress, the architectonics of the ground floor alone imply a very different organization of space than in fortresses of the eighth century B.C. Imperial period fortresses were tightly integrated structures that linked various

   10. The exact position of the entryway has not been clearly recorded on any extant plans
of the site; however, Piotrovskii (1969: 138) points generally to the southern complex just

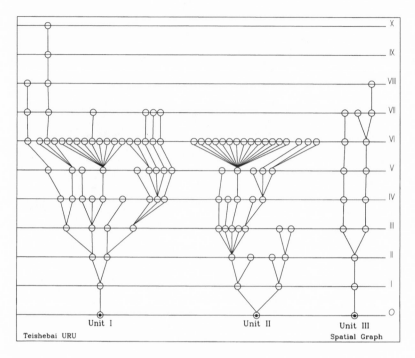

FIGURE 46. A spatial graph of Teishebai URU. (Source: author.)

institutions—religious, royal, distributive—into a single coherent imperial package, but the architectonics at Teishebai URU depart significantly from this model.

For example, the tight integration of components seen at Erebuni and
Argishtihinili does not appear to have been carried over into the layout
of Teishebai URU. The ground floor at the site was not integrated into a
single architectonic complex but was divided into a number of independent units accessible only from distinct points in the floor above.
Unit I, accessed from space A, comprises rooms 1–53; Unit II, accessed
from space B, comprises rooms 54–92; Unit III, accessed through space
C, is incomplete but includes rooms 96, 99–116. The RRA value for
Teishebai URU as a whole is 1.78, considerably higher than the more symmetric and open architectonics of Erebuni and Argishtihinili. Statistical

adjacent to the main gateway, and the local topography in this area does indicate a significant
rise against the wall face. Indeed, the presence of an entry ramp here may explain the lack
of a large tower on this wall face.

comparison of RRA values for the imperial period sites and Teishebai URU indicates a highly significant decrease in integration (an increase in asymmetry) in the reconstruction period regional center.[11] This division and disintegration of space suggests an increasing differentiation of the institutional components of the Urartian regime.

This impression of disintegration among the components of Urartian administration is reinforced by the redundancy of various functional parts within the fortress. The range of practices on the ground floor is generally limited to storage—wine and grain, dishes and utensils—and production—of beer, sesame oil, and bone and metal objects (Piotrovskii 1969: 139). Unlike the unified storage facilities at Erebuni and Argishtihinili, at Teishebai URU the excavators found several large, architectonically isolated rooms for storing grain (rooms 47, 58–64, 66–74) and wine (rooms 18–20, 23–27) accessible from very distinct locations on the second story.

Separation of redundant storage facilities in light of the overall disintegration of built space at Teishebai URU strongly suggests that the institutional components of Urartian authority had differentiated such that the administration of their assets was no longer fully coordinated. Furthermore, control values identify a handful of key spaces in each unit (rooms 10, 18, 78a, 104) that regulated movement throughout. Although spaces in Units I and II are fairly distributed, those of Unit III are highly undistributed, analogous only to the royal apartment area of Erebuni.

Without a greater understanding of the second story, it is difficult to identify the specific nature of the institutional components of the Urartian regime that may have claimed separate spaces on the ground floor. However, given the distinct room blocks occupied in imperial period Urartian fortress sites, it seems plausible to suggest that the best candidates for such a division would be along the same lines—bureaucratic/royal, religious/temple, and distributive/economic—detected in eighth century B.C. fortresses.

In considering the collapse of Urartian authority in southern Transcaucasia following Rusa II's renaissance, there is intriguing archaeological evidence for political disintegration during the reconstruction period. The traditional explanation for the Urartian collapse—conquest by Scythians—certainly begs the question as to why, less than a century earlier, Urartu had been able to weather Cimmerian and Assyrian attacks and reconsolidate, whereas military defeat in the late seventh century B.C. led to a

11. The difference between imperial period RRA values and those of Teishebai URU is highly significant ($p < 0.001$).

political collapse so complete that even the memory of the Urartian empire virtually disappeared. The spatial evidence for institutional differentiation at Teishebai URU suggests that the seeds of political collapse may have been planted by Urartu's own political strategies. It seems apparent that the imperial period Urartian polity in the Ararat plain was a coherent, well-integrated package that unified political, religious, military, and administrative spheres into a single regime. If, in response to the political crisis at the end of the imperial period, this structure was fragmented, allowing competing factions to assert power at the expense of the coherence of the whole, then this internal differentiation may go far to help us understand the institutional rivalries that prevented Urartu from successfully resisting Scythian incursions or reconsolidating after military defeat.

The experience of Urartian institutional landscapes suggests a historical process of fragmentation that ran counter to the increasing systematicity of perceptual aesthetics. Clearly, rivalries among institutions had a profound impact on built space, but did they also percolate into differing imagined landscapes? We have already discussed the representations of the Urartian political landscape provided by declarative inscriptions. Although these discourses on the spatiality of the polity can be attributed directly to the king, Urartian pictorial images of built environments are more ambiguous in their source, opening room for an account of how distinct imagined landscapes mediated institutional relationships.

IMAGINATION

Although the epigraphic representations of the Urartian political landscape discussed in chapter 4 ranged in their topologic detail from fields and vineyards to houses and granaries, Urartian pictorial representations of the built environment dwelt almost exclusively on rendering the built form of the fortress.

The large majority of Urartian art recovered to date can be classified as produced by the governing regime (Piotrovskii 1967: 15; Van Loon 1966: 166). Although little is known about the relations between political authorities and the artisans who accomplished artistic production in Urartu, well-provenanced representations of the built environment have been recovered from only two archaeological contexts—fortresses and tombs. Materials from fortresses can be rather directly attributed to the productive practices of the ruling regime, but materials from tombs are somewhat more problematic. Maurits Van Loon (1966: 166) suggests that the corpus of Urartian art may be divided into two groups: a court style directly connected with the Urartian governmental apparatus, and a popular style more closely linked with "commoners or provincials." In a gross sense, Van Loon separates these two styles by provenance, the popular style primarily represented by bronze belts from tombs of people presumed to be unconnected with the political apparatus, and the court style tied to fortress contexts. This would seem to require that we exclude images of the built environment known from bronze belts from a discussion of political art. But in a more subtle definition of his terms, Van Loon (ibid., 166, 168–69) defines the court style on the basis of its preference for straight horizontal and vertical lines, defining rectangular panels over the diagonal networks of lozenge-shaped sections typical of the popular style. Although several belts accomplished in the rectilinear court style contain representation of fortresses, no representations of the built environment are known from artifacts accomplished in the popular style. We are thus on relatively safe ground in treating the extant corpus of Urartian representations of the built environment as more closely associated with the governing regime than any other domain of Urartian sociopolitical organization.

It is very difficult to assess what portion of the total corpus of Urartian art includes representations of the built environment. These images are relatively rare on belts but somewhat less so on bronze plaques. Overall, the portion of the corpus devoted to such representations is probably small, though by no means insignificant. The following analysis is based on 34 distinct artifacts, which, though not exhaustive of known Urartian objects bearing representations of built environments, do include the large majority of the most complete and relevant pieces. (See Smith 2000: tables 1–3 for a full listing of the extant corpus of artifacts.)

Three compositional types of representations of the built environment can be identified in Urartian art: fortress elements, fortress images, and fortress scenes. Fortress elements appear in various media but share a common focus on stylized towers and abstracted stepped crenellations. Incised on the interiors of several bronze bowls from Teishebai URU (fig. 47), a stylized tree rises from the crenellated battlements of a simply rendered fortified tower.[12] Several of the bowls also included a short cuneiform inscription, encircling the tower, that specified the monarch to whom they belonged. This tower and tree motif also appears stamped into the handles of ceramic vessels, leading A. A. Vajman (1978: 104) to suggest that the symbol may have denoted "fortress" as part of the as-yet-undeciphered Urartian system of hieroglyphics.[13]

12. For a discussion of the iconography of the sacred tree in Urartu, see Belli 1980.

13. This is an interpretation accepted by A. E. Movsisyan (1998: 78) in his broader examination of Urartian hieroglyphs.

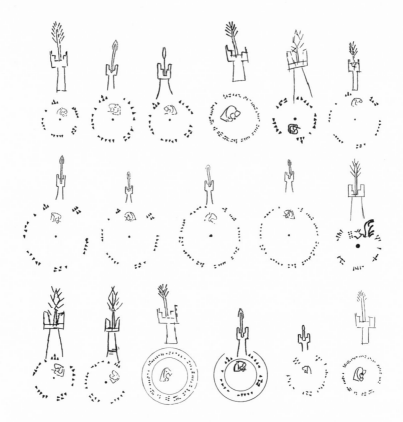

FIGURE 47. A selection of fortress elements on Urartian bronze bowls from Teishebai URU, with accompanying cuneiform inscriptions. (Redrawn from Piotrovskii 1955: figs. 28–32.)

The close association between the stylized representation of fortress elements and the political apparatus is further supported by the appearance of crenellation motifs in wall paintings recovered from Erebuni and Altintepe (Oganesian 1973: 45; Özgüç 1966: figs. 28, 29). Like Assyrian examples from Nimrud and Khorsabad, Urartian wall paintings were divided into "orderly panels and framed friezes" in which repeated motifs were regularly distributed (Azarpay 1968: 21). Fragments of painted plaster recovered at Erebuni indicate that a register was composed of a row of repeated crenellations set atop a zigzag-decorated cornice supported by projecting beams (Oganesian 1973: fig. 18). Rendered in black, blue, and white, the painted architectural elements at Erebuni appeared on the back wall of a columned hall thought to have been the throne room.

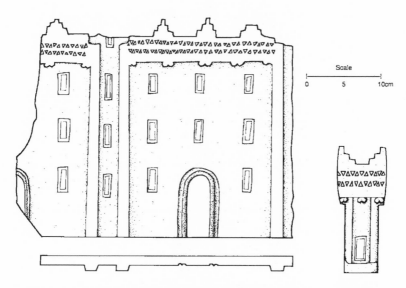

FIGURE 48. Two pieces of an Urartian bronze fortress model from Toprakkale. (Currently in the collections of the British Museum; drawing by author.)

In these two examples of the use of fortress elements, a very simple substitution is advanced; the most distinctive architectural dimensions of the Urartian fortress are mustered to give a tangible aspect to political relations. The political apparatus is considered from the perspective of its most prominent architectural signature.

Representations of fortress elements have been documented across the empire and in a variety of media, including a carved stone from Bastam (Kleiss 1974: fig. 8), a seal from Toprakkale (Wartke 1993: fig. 89), and an ivory fragment from Altintepe (Özgüç 1966: 89). Although these fortress elements do suggest that representations of the built environment were deployed to create associations between the regime and built elements, they tell us little about the nature of the relationship established. For that we must turn to more detailed renderings.

Urartian fortress images, the second class of pictorial representations of the built environment, provide considerably elaborated renderings of built facades, portrayed in strict frontal elevation, without compositional links to an explicit visual narrative. The best known of these representations is a bronze fortress model, now in the British Museum, recovered from the site of Toprakkale, near Van (fig. 48; Barnett 1950: 5–6). Only two fragments of what was likely a much larger composition now sur-

FIGURE 49. A fragment of an Urartian bronze belt. (Photo courtesy of the Archäologische Staatssammlung, Munich.)

vive. The largest piece represents a substantial segment of a fortress's lower facade, including the recessed curtine (the portion of wall between two buttresses, bastions, or gates) and projecting tower footings. The second piece is part of a large projecting upper tower that was once fixed atop the lower facade. The architectural elements of the model included crenellated battlements, projecting towers, and zigzag cornice decoration (marked by a double row of triangular slots that may have held colored inlays). In addition, we can also see an arched gateway (which seems to have been repeated in the adjacent curtine) and three rows of rectangular windows.

Fortress images are also known from bronze belts, though few have secure archaeological provenances. In one belt fragment, a fortress is shown set between double rows of embossed and engraved circles that are in turn flanked by a double row of hillocks incised with a scale pattern (Erzen 1988: 32, pl. XXXIV; Kleiss 1982: Abb. 1b). The fortress itself is composed of six projecting towers flanking five recessed curtines. Each panel of the curtine is surmounted by a zigzag cornice, whereas the towers are capped by both a cornice and stepped crenellations. An arched double-winged gateway, with its right door closed and left door open, occupies the center curtine panel; one large rectangular window opens from each of the other panels. Small embossed circles and inscribed triangles surround the composition, giving a sense of elevated topography to the entire image.

Another belt fragment (fig. 49), currently in the collections of the Museum of Anatolian Civilizations in Ankara, Turkey, portrays a fortress, flanked by panels depicting parades of griffins and fish. The fortress itself is composed of two towers projecting from a curtine wall. Atop both curtine and tower are the now-familiar zigzag-decorated cornice and stepped crenellations. Rectangular windows open from both the curtine and tower. Off center in the curtine is an arched double-winged gateway with its left half closed. This structure is set off from other panels by a row of embossed circles on each side. In the panel to the left is a parade of four winged creatures with hooked beaks that resemble griffins. In the panel to the right are three fish, positioned end to end.

Only a limited repertoire of figures and scenes are depicted surrounding the fortress in these contexts. In the belts cataloged by Hans Kellner (1991: nos. 255, 261, 269, 279, 282) we can see flanking panels with images of:

· fish; a winged beast; two figures on both sides of a large jar;

· fish; winged beasts; birds; sheep; a figure (perhaps a deity) on a throne;

- fish; birds; two figures, one holding a small spouted flask, flanking a large jar;
- fish; birds; sheep; two figures, one holding a spouted flask, flank a large jar; a partial scene with two standing figures bearing various items approaching a figure seated on a throne in front of an altar with dishes placed on it (the altar is surmounted by bulls' horns);
- sheep; goats; two panels of processing figures holding "offerings"; a large panel with two processing figures bearing "offerings" approaching an altar with dishes on it (the altar is again surmounted by a set of bull's horns); on the other side of the altar, a figure sits on a throne while being fanned by an attendant;
- In two long registers without division into separate panels, two fortresses anchor a long procession punctuated by what seem to be ritual scenes.

The animals surrounding the fortresses may contain geographic information linking the fortresses to specific places (such as geese and swans to perhaps signify Lakes Van or Sevan) but they also embed these constructions in the symbolism of the natural world. The primary activities depicted with the fortress images were religious rituals, often explicitly focused on what appear to be deities. However, only in the last example are the fortress and the ritual scenes explicitly conjoined, incorporating the fortress into the religious performance (Smith 2002). In most of the belts, panels separate the fortress from the other scenes, thus making it difficult to define the exact relationship between religious ritual and the built environment of the Urartian political apparatus that these images construct.

Each of the fortress images share several common architectural features: stepped crenellations atop the battlements; zigzag friezes on the cornice; and high, narrow towers projecting from recessed curtine walls. Wolfram Kleiss (1982) has suggested that these three architectural elements constitute the core repertoire of symbolic elements in Urartian fortress representations.[14] Because of this repeated core repertoire, Kleiss cogently argues that these images portray Urartian fortresses rather than those of rival polities. The core elements are complemented by a more variable set of details, such as rectangular windows in the curtine wall and tower base and arched, double-winged gateways (often with one side open).

---

14. He points out (Kleiss 1982: 54) that the zigzag friezes depicted on the cornices are a distinctly Urartian artistic element, unknown elsewhere in contemporary southwest Asia.

Kleiss suggests that Urartian fortress images were symbolic expressions of the strength of the ruling regime (ibid., 54). However, as Peter Calmeyer (1991: 316) points out, the depiction of fortresses with gateways half-open hardly promotes an image of them as inviolable. The elements that accompany the fortress images on belts described above seem more intent on placing the fortress within a repertoire of natural symbols and ritual performances than advancing claims of the martial power of the king. Only in one known example, a belt fragment currently in the British Museum, do we see a fortress associated with warriors (suggesting that there may well have been media that advanced martial images of the built environment).[15] But the coercive power of the Urartian regime does not seem to be the dominant message conveyed in the large majority of fortress images. The use of stylized mountains and parades of griffins and fish inscribes the fortress within a larger context, proclaiming a specific relationship between a built form and the natural world. The emphasis on religious rituals embeds the fortress in a supernatural landscape, rather than a political one. By naturalizing (or supernaturalizing) the fortress, its political content was downplayed.

In contrast to the generally static composition of fortress images, fortress scenes used architectural elements as backgrounds for figures engaged in some form of action. The clearest, not to mention most monumental, known example of a fortress scene was found on several stone blocks from the site of Kefkalesi, on the northwestern shore of Lake Van (fig. 50; Bilgiç and Öğün 1967: 16–18). Each square stone block (approx. 1.40 × 1.40 × 1.10 meters) was carved with the same scene on all four sides, depicting the facade of a fortress being blessed by a symmetrical pair of winged deities astride lions.[16] The fortress is composed of three projecting towers flanking two recessed curtine walls. The curtine walls are largely obscured by the deities but do include five T-shaped windows placed around the figures. The battlements atop the curtine are set atop projecting support beams rendered in evenly spaced semicircular pairs. A doublerowed zigzag cornice surmounts the support beams and is itself topped by stepped crenellations. Perched atop the crenellations, a symmetrical

15. Warriors and hunters can be seen on a number of belts (including Kellner 1991: nos. 182, 175, 174, 117), so we know that these figures were a part of the Urartian representational palette.

16. Although each side of the block bears identical scenes, there are some small stylistic differences in each, raising the question as to whether this piece was accomplished by one or more artists. More detailed consideration of the carving of this important piece might reveal significant insights into the organization of Urartian artistic production.

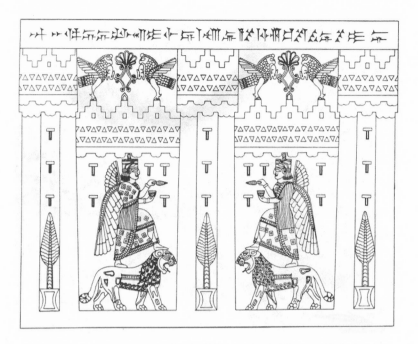

FIGURE 50. An Urartian stone block with carved relief from Kefkalesi. (From Ögün 1982.)

pair of winged monsters clutch rabbits in their beaks. The towers project 0.7 centimeter out from the curtine. They are divided into three vertical panels by a recessed center. Set in each of these recesses is the Urartian sacred tree—a thin vertical stalk set in rectangular planters with concave sides sheathed by a hatched lanceolate body—and three T-shaped windows. The battlements atop the tower are built on two rows of projecting support beams. The battlements include a zigzag cornice and stepped crenellations. Across the top of the stone is a fragmentary inscription that reads: "Rusas, son of Argishti, has built this place [E2.ashihusi],[17] thanks to the grandeur of god Khaldi. Its found[ation] was ill maintained . . . was not. I, Rusas, have built. No matter whoever destroys this inscription . . . whoever place . . . , and the things entrusted to me by god Khaldi, let him

17. Mirjo Salvini (1969: 14–15, personal communication 1998) suggests that Bilgiç and Öğün's translation of "E2.ashihusi" as "place of cult for drinking and sacrifice" is incorrect. The correct translation is not known but likely refers to a building part close to the fortress storage rooms.

be annihilated by Utu (god of the Sun)" (Bilgiç and Öğün 1967: 18; Salvini 1969: 14–15).

The symmetrical figures in front of the fortress appear to be supernatural figures, perhaps the god Khaldi or more minor protective deities or genies (Burney 1993: 108; Seidl 1993: 559; Wartke 1993: 66). The lions on which the figures stand are shown striding toward each other—toward the center of the composition—and are decorated with a scale-like motif. Each figure, with one foot on the lion's back and one on its head, is garbed in a decorated skirt-like garment extending to the ankle on the back leg and mid-thigh on the forward leg. On their head, each wears a crown with horns on each side, topped by a disk set on double volutes. The wings of the supernatural figures, marked by feather-like hatching, are attached to the figures' backs, with the front wing extending down and the back wing protruding up to suggest compositional depth. Each figure holds a cup in the left hand and the Urartian sacred tree in the right (the use of particular hands must be of ritual significance because the artist compromised the symmetry of the piece to keep the bowl and tree in the left and right hands, respectively).

The narrative of this fortress scene appears to be devotional. The god(s) appear astride a lion, offering their blessing and their protection. Not only does this transform the fortress—an explicitly political locale—into a place infused with religious significance but it also offers up the activities of the political apparatus as ritual activity. The primary figurative process at work in the image represents the relation between the political apparatus and the cosmos as mediated by the beatifications of the deity.

Fortress scenes assembled on similar figurative lines also appear in a host of bronze plaques. Although it is useful to consider the general compositional elements of these artifacts, only a small number, such as the plaque recovered at the site of Erebuni (fig. 51), boast a well-described provenance (Khodzhash, Trukhtanova, and Oganesian 1979).[18] As a result, the sizeable corpus of plaques is not sufficiently robust in its archaeological detail to support high-resolution stylistic analyses. However, the fortress scenes rendered on these plaques are worth discussing on a general level because they deploy figurative elements very similar to those rendered in the Kefokalesi reliefs and the fortress images discussed above. The overall composition of the plaques is significant in that they suggest some attention to foreground and background. The spatial convention used in the scenes

18. The plaque from Erebuni was recovered during excavations of room D in the Urartian settlement outside the fortress walls.

FIGURE 51. An Urartian bronze plaque from
Erebuni. (Erebuni Museum #15/32/AB/71.
Photo courtesy of the Institute of Archaeology
and Ethnography, Republic of Armenia.)

places a zigzag cornice and stepped crenellation along the top border of
the composition, thus tying the scene to the built environment of the
fortress. The architectural elements rendered in plaques are highly stylized
but recognizable as part of Kleiss's core elements of fortress representation. The crenellations are always two-tiered and are typically either inscribed or cut out of the top of the plaque, as in the case of the plaque from
Erebuni. A cornice is a more variable feature that is often left out of the
composition. Cornices are embossed or incised. Despite their generally stylized form, both the crenellation and the cornice (when included) are still
recognizable as elements of the core repertoire of fortress features that were
mustered individually as fortress elements or as details of fortress images.

The figures that constitute the narrative content of the scene occupy
the remaining lower space of the composition. The scenes presented in
the plaques show deities and suppliants engaged in devotional ritual. Although all of the scenes in these plaques carry a nominal sense of action—

a devotional ritual—what appears to be of greatest significance was simply the portrayal of the deity in front of fortress walls. Indeed, in many examples the supplicant, whose entreaties would presumably be the narrative focus, is left out of the composition.

In looking at all three classes of pictorial representations—fortress elements, fortress images, and fortress scenes—we can outline a narrative on political legitimacy that differed significantly in its aesthetic appeal and ideological content from that detailed in the epigraphic sources discussed in chapter 4. This narrative has three dominant elements: extension, reduction, and integration.

Extension. In the deployment of fortress elements, the incorporeal entity of the regime is lent a materiality in the shorthand form of architectural elements of the fortress, suggesting that Urartian authority can be apprehended in relation to its built environment. Given the prominence Urartian fortresses would have had on the physical landscape, it is unsurprising that their form might hold, from a phenomenological point of view, considerable poetic significance.

Reduction. Fortress images and scenes situated the imperial apparatus within a broader discursive field. In so doing, the political apparatus was reduced, nameable in reference to one part of its domain. In fortress elements and fortress scenes, the primary locus of imperial power was removed from the particular time and place of its construction by rendering only the most generalized sense of the built environment—bereft of any historical specificity and only a very general sense of location (possibly keyed by the animals surrounding fortress images). The semiotic potency of the fortress to stand in for the broad set of relations and referents that defined the political apparatus, as revealed by the glyphic use of architectural elements alone, lent it a remarkable portability.

Integration. The realization of the political ideology carried in pictorial representations of the built environment can be seen in both fortress images and scenes. The fortress is removed from the political domain and sacralized. Fortresses are not rendered as sites of war, bloodshed, and domination; they are places of the deities. It is through integration of the part (fortress) with the whole (the transcendent world of the deities) that Urartian pictorial representations of the built environment make their sternest arguments for the legitimacy of the political order in the imagination of the political landscape.

• • •

What is fundamental to understanding the use of these figurative processes to secure legitimacy is an identification of the transcendent principles presumed to link the fortress with devotional ritual. The legitimacy of the Urartian regime flowed, these scenes seem to argue, not from the tectonic charisma of the king but from supernatural sources, specifically the beatification of the gods. The fortress was removed from its specific political and historical circumstances and transformed into a sacred place, apolitical, ahistorical, and depersonalized. Although the textual representations of the built environment described in chapter 4 emphasized the historical specificity of the landscape and its emergence at the hands of the conquering king, pictorial representations of the built environment portrayed the created environment as a transcendent, ahistorical, and de-politicized site of devotion and blessing. One appealed to a triumphal aesthetic of conquest; the other, to a transcendent aesthetic of the sacred.

Ultimately we must pose the question: to what can we attribute this division in the aesthetics of two representational programs that sought to define the imagination of the Urartian political landscape? Three possible interpretations come to mind. The first looks to differences in the media carrying each program. Such an account would focus on inherent formal possibilities and restrictions within the two modes of discourse — such as the difficulty in presenting a depersonalized political apparatus in word. This is not a compelling account, because we know from contemporary Assyrian contexts that pictorial media certainly lend themselves to carrying the charismatic claims to legitimacy that we find only in Urartian epigraphic sources (see Winter 1983: 24).

Second, it is possible to account for varying ideological programs in reference to differing constituencies to which they were directed. It must be admitted that identification of the audience for written and pictorial representations is perhaps the most problematic issue for any study of cultural production in early complex polities (Michalowski 1990, 1994; Postgate, Wang, and Wilkinson 1995). In order to embrace this interpretation, we must assume that the epigraphic and pictorial programs were accessible by distinct segments of the general populace. However, both programs seem to have been directed primarily to an elite. Pictorial representations are best known from within Urartian fortresses and thus would have been viewed only by those allowed within it walls, an undoubtedly restricted corps of imperial officials, foreign emissaries, local governors, and courtiers. Literacy in Urartu was likely restricted to an elite scribal class, severely limiting the number of people who could directly read the words of the king. However, problems of dissemination of information carried in word and image are easily overcome, either through public recitations or, in the case of images, by opening rituals to the public.[19]

The third, and to my mind most compelling, interpretation looks to different institutional sources of ideological production within the Urartian regime. What seems clear is that the epigraphic messages discussed in chapter 4 come directly from the monarch. However, the pictorial representations described here are not so directly attributable to the king. Indeed, many pictorial representations are more directly associated with religious than with royal contexts. In this interpretation, the differing ideological programs for securing legitimacy are the products of distinct institutions within the governing regime seeking legitimacy for Urartian political authority in terms most favorable to its factional status.

It should be noted that a distinction between royal programs in word and temple programs in image cannot be overdrawn. The royal inscription on the Kefkalesi stones and the use of fortress elements in the royal receiving hall at Erebuni mitigate against drawing solid institutional boundaries between these two programs. Furthermore, I do not want to give the impression that Urartian inscriptions do not deal with religious matters. In fact, many Urartian texts contain discussions of religious beliefs, rituals, and the pantheon.[20] As a further caution, the institutions of king and temple should not be conceptually disarticulated, a point emphasized by a 1989 study of the links between King Ishpuini and the emergence of the cult of Khaldi as a state religion (Salvini 1989: 88–89).

The suggestion that the dominant modes of political representation are not always unified, singular, and coherent adds some complexity to investigations of imagined landscapes; statements of various institutionally located representational programs may or may not be expressive of the regime as a whole. Hence, to ignore relationships among political institutions and their role in constituting authority is not only to lose sight of a critical apparatus of governance, as March and Olsen have argued quite

19. John M. Russell (1991: 238–39) notes that Assyrian texts from the reigns of Assurnasirpal, Sargon II, Sennacherib, and Esarhaddon refer to festival occasions when palaces were opened to a more general audience of Assyrian subjects.

20. The Meher Kapısı Monument (König 1955; Salvini 1994) is a particularly important source for interpretations of the Urartian pantheon. For more general discussions of Urartian religion, see Hmayakian 1990; Salvini 1989.

strongly in political science, but also to miss the profound linkages and rivalries that distinguish prominent places of sociocultural production.

Authority and the Institutional Landscape

The description of the institutional landscape of Urartu presented above presumes a theory of architecture that connects buildings to politics not just through evocative aesthetics but also through the experience and imagination of landscapes. Le Corbusier (1970: 211) pointed toward an encompassing theory of the politics of built environments with his slogan "Architecture or Revolution. Revolution can be avoided." Although overstating the direct power of the architect to intercede in contemporary struggles, Le Corbusier firmly established High Modernism's sense of social commitment to political transformation through the medium of built form. The vision of modernism as architecture for the masses, as a space of social transformation that might break down traditional divisions of class or gender, provided a highly empowered position for design as well as for architectural interpretation. If buildings were not just passive expressions of a dominant zeitgeist, then they could be understood as more profoundly instrumental in directing and securing the reproduction of existing sociopolitical formations and promoting new imaginings of future forms of social life and governance. However, modernism's political vision eventually became clouded by accusations of complicity, not just in the totalitarian architectures of Nazi Germany and Stalinist Russia (contexts where a spurning of modernist aesthetics were combined with enthusiasm for its theorization of built power) but also in the dehumanizing effects of technological space over meaningful community (as displayed in "the desolation of mass housing projects, the wasteland of urban renewal, [and] the alienation resulting from an architectural language that now seemed arcane" [McLeod 1998: 683]).

Postmodernism arose in part from a profound disillusionment with the complicity of the politicized architect in designing spaces of twentieth-century violence, poverty, ecological degradation, and colonialism. Such tragedies of the modern contributed to a shift in the priorities of interpretation and design away from architecture's social role and political sources (Frampton 1992; Jencks 1973, 1991; Venturi 1966). Postmodern aesthetics arose as an attempt to strip away its predecessor's focus on the transformative power of architectural production to direct social action and replace it with a concern for the meaning-laden contexts of its consumption. As a result, postmodern architecture, to the degree that a unified account can be assembled of a highly plural movement, has come to refocus design and interpretation away from metanarratives of liberation, reason, and equality and toward aesthetics that forthrightly engage historical allusions, pop styles, and regional variation in modes and media of expression (McLeod 1998; Stern 1980). At its best, postmodernism has developed a form of negotiated accommodation to corporate demands through a preservationist impulse that attempts to rebuild and revive communities rather than cloistering them into high-rises. However, at its worst, postmodernism has come to stand for a complete withdrawal of architecture from the world or, perhaps more precisely, the insulation of the architect from the social. For example, Peter Eisenman (1984: 166) describes his goal as "architecture as independent discourse, free of external values," whereas Bernard Tschumi (1987: viii) has described an architecture "that means nothing." In effecting such a complete withdrawal from theorizing the articulation of architecture and politics, postmodernism has come to be derided as the "architecture of Reaganism," an apotheosis in built form of Western consumer culture where history is reduced to nostalgia, communities refigured as prefab simulacra, and aesthetics restricted to superficial ornamentation (McLeod 1998: 680).

Architectural theory thus finds itself in a difficult position in its understanding of politics, caught between modernism's doomed faith in the political potency of formal production and postmodernism's ostrich-like attempts to insulate itself from politics by emphasizing the playful neutrality of surface and image.[21] What is underdeveloped in the debate over design is a historical theorization of built environments in practice, an anthropology of architecture that can link both the interest in production with an account of use. This is where the archaeology of early complex polities can present an account of architecture within larger political landscapes that grasps for the political interests in the architectonics of form, the evocative aesthetics of surfaces, and the imaginings of proper (built) orders. Such a project can begin with explicitly political spaces. In the years since World War II, the federal government of the United States has become one of the world's largest builders in terms of architectural volume and one of its most omnipresent landlords, with buildings in most countries of the world. It has been plausibly suggested that one might traverse the entire span of the continental United States without ever leaving built

21. See Hal Foster's (1985) critique of postmodern politics.

and unbuilt environments created by federally commissioned architects and engineers (Lacy 1978: vii). And yet, despite the ubiquity of explicitly governmental architecture, the sources and consequences of form and aesthetics remain woefully undertheorized.

The picture that emerges from the foregoing account of institutional relationships is one of historically embedded sites of production of the experience, perception, and imagination of Urartian political landscapes. As such, the architecture of Urartian institutions was constitutive of negotiations over sectional visions of political power and legitimacy. Architecture here emerges as a craft of authority, in the sense that Susan Kus and Victor Raharijaona (2000: 110) use that term in their study of Imerina politics in nineteenth century A.D. Madagascar. Moving deftly from the domestic to the institutional, Kus and Raharijaona describe the parallel "crafting" of a royal capital and a unified polity, both of which were founded on the reiteration of tropes of domesticity and the cosmos in landscapes. Their study of the building of the Imerina polity in the production of the landscape of regime and the institution of kingship foregrounds the constitution of authority within relationships that range rather freely across scales. The result is that we must turn back once again to reconsider each of the relationships constituting political authority that have been examined in preceding chapters.

Conclusion

Toward a Cartography of Political Landscapes

Just as none of us is outside or beyond geography, none of us is completely free from the struggle over geography. That struggle is complex and interesting because it is not only about soldiers and cannons but also about ideas, about forms, about images and imaginings.

Edward Said, *Culture and Imperialism*

Above a gate into his imperial city of Samarkand, the legendary Timur ordered inscribed the resounding architectonic boast "If you doubt our might—look at our buildings" (Kapuściński 1994: 77–78). Although their straightforward syntax clearly denote a politically motivated concern for the built environment, these lines conceal as much about the relationship between landscape and political authority as they reveal. We might interpret Timur's magniloquence in a number of ways. One would be to translate the volume, energetics, or size of Samarkand's built environment into a figure that might be compared with that of, for example, Persepolis or Anyang. In that way we might come to a relative assessment of Timur's might by ranking Samarkand along a general historical scale of political centers, assessing this much-feared oasis polity's development in relation to other early States. But in making such an interpretive move—an absolutist turn—we ultimately fail to heed Timur's call to attend to the buildings, because the absolutist vision must strip away all that was specific to the Timurid capital in order to bolster the reductionism of its foundational social evolutionary project.

A second interpretation of Timur's inscription would describe the architecture of fifteenth century A.D. Samarkand as an evocative memorial to the conqueror, a sublime portrait in brick and tile of the martial glory of the desert king. This was certainly the way Christopher Marlowe enshrined the oasis-city in his epic to Tamerlane:

> Then shall my native city, Samarcanda . . .
> The pride and beauty of her princely seat,
> Be famous through the furthest continents,
> For there my palace royal shall be placed,
> Whose shining turrets shall dismay the heavens,
> And cast the fame of Ilion's tower to hell.
>
> (1999: IV.iii. 107, 109–13)

Yet Samarkand is no Troy, neither a massive fortress nor a testimonial to the martial spirit that lent Timur's armies a reputation for building towers and walls from the skulls of the vanquished. Ryszard Kapuściński describes Samarkand as "inspired, abstract, lofty, and beautiful; it is a city of concentration and reflection" (1994: 77). Indeed, the lack of rampant militarism in the design and organization of the city has led one scholar to question whether Samarkand, a city that encourages reflection, mysticism, and contemplation, could have been built by such a ruthless marauder (Papworth cited in Kapuściński 1994:78). Although there can be little doubt that Samarkand bloomed as the capital of central Asia under Timur's tutelage, the juxtaposition of his brutality and Samarkand's gentility provides a valuable caution against forms of spatial subjectivism that essentialize relationships between political organization and landscape aesthetics. A satisfactory account of Timur's Samarkand must describe how authority was constituted through landscapes that simultaneously inspired and terrorized, that disciplined bodies even as it turned minds to the heavens.

Thus, a third, more encompassing, interpretation of Timur's inscription would approach both the built environment of Samarkand and the king's inscribed representation as constitutive elements of imperial politics—a landscape that simultaneously constituted, and was itself constituted by, Timurid authority. Samarkand's buildings do provide us with a sense of Timur's might, not as an absolute index of social evolutionary development or a general lexicon of human expressive aesthetics but as an instrument for establishing physical, expressive, and imagined political relationships. It is this perspective on the articulation of landscapes and political authority that has occupied the preceding pages.

As a conclusion to these investigations, I want to address three issues crucial to the intellectual future of early complex polities in social science research: the role of comparative analysis; the articulation of contemporary political projects with studies of ancient landscapes; and the integrity of early complex polities as objects of investigation.

Constellations and Comparison

In recent years, studies of early complex polities have proven exceedingly reticent to move beyond single cases to develop models and theory within comparative analyses. Gil Stein (1998: 2) has lamented the paucity of comparative accounts of complexity, noting that the results of such geographic restrictions are a diminishment of cross-regional dialogues and a certain hesitation when moving from empirical studies to theoretical argumentation. To these complaints we might add the growing intellectual separation of the theoretical priorities orienting research design and those that frame interpretive writing. That is, although field investigations of early complex societies have in significant measure moved to position research within a post-social evolutionary intellectual climate, where issues of power, ideology, and agency have supplanted formal typologies and meta-narratives, there have been few attempts to describe how these alternative conceptual threads might interweave interpretations of discrete cases. One important exception is the call by John Baines and Norman Yoffee to develop comparative accounts of ancient "civilizations" in reference to the interrelated conceptual locations provided by order, legitimacy, and wealth. For Baines and Yoffee, these three concepts provide a basis for interpreting the apparatus of "high culture": "Through an investigation of the instrumental principles of hierarchization, of the restriction and display of certain kinds of wealth and the devaluation of other kinds of symbols, and of constitutive institutions of legitimation that emphasize the dispersive ties between rulers and ruled we argue that a comparative method can be pursued" (2000: 13, see also 1998). Like this investigation, Baines and Yoffee root the impetus to comparison not in the parity of specific social evolutionary forms but in the operation of sociopolitical life.

In adding to their call for renewed comparative study of early complex polities, I suggest that two priorities must guide the formulation of a new representational vision. First, the spatial reductionism that allowed social evolutionism to compress variation must be emphatically resisted; it is the points of variability that are the very things comparative study should bring out if we are to develop a truly anthropological vision of early social life. Second, we must resist the insistence on complete complementarity in our objects of investigation. Such equivalences were always products of theory that allowed us to strip away variability rather than any real essentials of human social life that might provide inherently stable foundations for moving across cases. As a result, a comparative account of early complex societies must break traditional formal boundaries to find illumination wherever it might be found.

Although in one sense this approach might be understood as simply opportunistic, it is, in a more programmatic sense, constellatory, to use Walter Benjamin's term: "Ideas are not represented in themselves, but solely and exclusively in an arrangement of concrete elements in the concept: as the configuration of these elements. . . . Ideas are to objects as constellations are to stars" (1977: 34). The goal of comparative investigation is not the uncovering of the essential idea of history that lies behind archaeological and epigraphic phenomena, but rather the illumination of conceptual linkages embedded in the specificity of our objects. A constellatory approach to comparative investigations of early complex polities steadfastly refuses to seek interpretive refuge in a foundational historical metaphysics, either social evolutionary or historicist, breaking with broad totalizing intellectual projects without descending into mere empiricism. It is important that, in developing a constellatory approach to the comparative anthropology of early complex polities, we maintain the ability to move between traditional horizons of objective and subjective analyses (see Adorno 1977; Eagleton 1990: 332–33).

It is in refusing all allegiances to such outmoded commitments that investigations can provide not only productive accounts of social life but also novel epistemological visions. Such efforts to erect knowledge of the past atop multiple foundations offer a way out of the moribund conflicts between positivism and interpretivism that have bled archaeology and other historical social sciences of so much of their theoretical imagination. Such an intellectual movement also points the way toward important and interesting rapprochement between archaeology, art history, and epigraphy—spheres of expertise that have tended to guard disciplinary prerogatives through exclusionary epistemological procedures rather than welcome overlapping interpretive possibilities. But it is just these hybridized forms of understanding the past that will allow early complex polities to speak in meaningful ways to contemporary political problems.

Ancient Landscapes and Contemporary Politics

Concurrent with the declining appetite for comparative investigations since the social evolutionary heyday has come a general uncertainty as to the contemporary political project that underlies the study of early complex polities. That is, where vigorous debates over the primacy of coercion and consent or circumscription and class in the origins of the archaic State might be quite clearly understood as proxy battles in Cold War historiography, what can the examination of the spatial production of authority in early complex polities tell us about political life today? There are a number of answers to this question. First, authority has locations. It produces itself in space and thus has a real physical position. But this spatiality of authority is highly discontinuous. This means that critique must be embedded in place: how do spatial forms, aesthetics, and images intersect to realize power and secure legitimacy? As buildings, streets, and pathways are cut, as cities grow and shrink, how do new forms and the reproduction of old ones sustain or threaten existing regimes? Second, in imagining new political possibilities, we must envision the spaces that they would depend on and the consequences of these spaces for those who lived there. It is this critical project that provides a point of articulation between archaeological interpretation and political thought.

As should be clear, this book is as much about the way we approach contemporary politics as it is about the study of Urartian institutions or Classic Maya geopolitics. But what does a cartography of political landscapes imply for issues of contemporary policy? Few studies of contemporary politics have taken landscape seriously as a critical element in the production of civil communities. One exception to this is *Laws of the Landscape: How Policies Shape Cities in Europe and America* (1999), Pietro Nivola's comparative study of the effects of political policies on the physical shape of cities in Europe and America. Despite the work's subtitle, Nivola appears to be rather skeptical about the potential effect of political decisions on modern urban landscapes and social life. Although proposing various ways to raise funds to improve school performance and reduce urban crime, Nivola is largely dismissive of the Clintonian remedies that have become the basis for current approaches to urban problems, such as enterprise zones and tax incentives—policies explicitly directed toward re-mapping the experience, perception, and imagination of inner-city landscapes. Nivola's analysis leaves the American reader in a quandary. On the one hand, Nivola echoes the desire for livable communities based on the European model (Paris, London) held together less by private automobile traffic than by easy walking distances and public transport (the community aesthetic of both James Kunstler [1993] and the New Urbanism). On the other hand, Nivola appears convinced that the politics that produce such livable cities inevitably result in higher unemployment and the various follow-on problems that currently grip urban centers such as Paris and Rome.

Nivola's study illustrates a central problem in the current, largely implicit, discussion of the political landscape: a deeply held conviction that political decisions can only address the margins of contemporary problems rooted in the spatiality of our current world. The big decisions—those with the potential to truly reshape cities, small communities, and rural farmland—are made by private corporations and "the market." For Nivola, this seems to be a positive development; the American city exemplifies the power of economic interests to shape places that, though perhaps not beautiful or livable in the European sense, do produce relatively high levels of employment. For Joseph Rykwert (2000), this corporatization of American communities (most dramatically illustrated by Disney's planned community of Celebration, near Orlando, Florida) is a deplorable development because it places profit ahead of civic interests in the planning and design of buildings and cities. Rykwert agitates for more pronounced public involvement and a limitation on the power of exclusively business interests to shape the places in which we live.

Both Nivola's and Rykwert's analyses reflect a serious failure of the modern political imagination that is rooted firmly in the modernist refusal to adequately theorize the profound links between landscape and politics. For both, a very limited sense of the political (restricted primarily to explicit policy interventions rather than to sets of civic relationships) leads to a deep cynicism about the power of governmental decisions to shape any place. Yet political decisions are what open room for corporate assumption of the power to shape the landscape. Celebration, for example, was vetted through several regulatory bodies in the state of Florida before Disney was even allowed to clear the site. Public agencies have the authority to intervene in the construction of the American landscape at numerous points in the development process; however, the current array of governmental forces often makes the explicit decision not to do so. This decision to defer to business interests is itself a political decision made in an increasingly corporate-centered political climate, where coalitions and financial support for rival claimants to the levers of power are predicated more on market values than on civic ones. Indeed, communities such as Celebration rest on a well-developed argument that corporate values are the same as civic values, a preposterous idea that nevertheless has confined our imagination of the proper operation of the political within contemporary communities (Frank 2000). The problem is thus not with a limited set of options for intervening, as Nivola suggests; opportunities for intervention abound. The problem is in our very limited imagination as to where, how, and in what capacity the political should intervene.

It is, I hope, clear that the chapters in this book are not intended to be a libertarian blast against the encroachment of politics into the landscape; politics is always already a part of our landscape. The problem is not to exorcise it but rather to effectively and deliberately control how political authority is exercised in the landscape because that landscape is critical to the reproduction and transformation of civil society. We should care about the landscape not simply because it makes life more or less pleasant (in the expressive aesthetic of the New Urbanism movement) but also because it profoundly conditions the very terms in which we are situated as subjects, governed by institutions and regimes, and located within polities and a geopolitical order. The landscape is thus central to all facets of political authority, from global frameworks for peace and security to community efforts to mobilize coalitions.

"Early Complex Polities" Revisited

If we can promote visions of the past that inform politics in the present, how tenable is the term used to describe the object of this inquiry (the "early complex polity")? In his profound study of modern large-scale state-sponsored social planning, James Scott provides a rather convincing demonstration that what lies at the heart of most political programs is not complexity but rather a drive to simplicity. In his study of modernist agricultural policy, Scott emphasizes that politics can be understood as a failure of representation:

> The necessarily simple abstractions of large bureaucratic institutions, as we have seen, can never adequately represent the actual complexity of natural or social process. The categories that they employ are too coarse, too static, and too stylized to do justice to the world that they purport to describe. . . . [S]tate-sponsored high modernist agriculture has recourse to abstractions of the same order. . . . Unable to effectively represent the profusion and complexity of real farms and real fields, high modernist agriculture has often succeeded in radically simplifying those farms and fields so they can be more directly apprehended, controlled, and managed. I emphasize the *radical* simplification of agricultural high modernism because agriculture is, even in its most rudimentary Neolithic forms, inevitably a process of simplifying the profusion of nature. (1998: 262, emphasis in original)

Many of the political practices discussed in the preceding chapters quite clearly hinge on producing simplicity out of complexity, whether because of a failure to grasp the true intricacies of the situation or a strategic resolve to promote stark contrasts rather than shaded nuance.

If we are to understand early complex polities as, in fact, engaged in systematic efforts at simplification, then the very notion of political complexity appears not simply oxymoronic but truly misguided. As I noted in the introduction to this book, the designation "early complex polity" can only be provisional—a shorthand for objects whose relationship to each other lies not in an essential spatial or historical idea but only in the linkages established by the conceptual apparatus developed here to splice them together. Within the account of politics developed in these pages, it should be clear that, though a great deal is at stake in a re-conceptualization of politics in ancient worlds, very little rides on the specific terminology of categorical fragmentation. Thus, whether or not the cases discussed here hang together as early complex polities is far less significant than whether they hang together as political landscapes. This is not to advocate against the use of the term early complex polities but merely to problematize it as I have other categorical terms.

Prospects and Horizons

In bringing these discussions to a conclusion, I want to briefly note several of the key choices made here in negotiating the difficult theoretical terrain that lies between landscape and authority. Undoubtedly the decision most visible to archaeologists will be the modest analytical position that I have accorded the natural environment. The environment plays a rather small role in the preceding analyses relative to more traditional accounts—one more descriptive than causative. Although chapter 2 indicates a few of the more theoretical reasons for a shift in focus away from environment and ecology as embracing determinants, it is important to emphasize that, in arguing for an understanding of politics through landscapes, I am not arguing against the relevance of other factors, including the natural environment. Clearly, basic environmental parameters are critical to human life and have considerable impact on the production of landscapes. However, fluctuations in environments must be understood within the social and political ordering of daily practices, not exterior to them.

In a recent discussion of why the southern alluvium of Mesopotamia presented the ideal environment for the initial development of social complexity during the fourth millennium B.C., Guillermo Algaze chastised archaeologists of the past decade for neglecting the centrality of ecological conditions to social development. The profound transformations in the societies of southern Mesopotamia during the fourth millennium B.C., he writes,

> can only be understood against a background combining the extraordinary transportational advantages offered by the Tigris-Euphrates fluvial system, the unique density and variety of subsistence resources . . . and the absence from that environment of other necessary resources, such as metals, timber, and status-validating exotics. The synergy created by these conditions spurred the creation of high levels of social and economic differentiation, promoted unprecedented population agglomerations . . . and selected for the creation of new forms of social organization and technologies of social control. (2001b: 204)

Algaze places a tremendous historical burden on southern Mesopotamian ecology; however, each of the elements of the local environment that he highlights as determinative must have already been constituted within a sociopolitical field for them to have cultural salience in Mesopotamian historical transformations. That is, we must presume an existing set of political interests that vest status in exotic trade goods, that predicates political economy on a diverse base of subsistence resources, and that founds geopolitics on the exchange of bulk commodities (more effectively transported by river than land). Were it not for the promulgation of these political interests in specific features and capacities of the Mesopotamian natural environment, fourth millennium B.C. sociopolitical transformations would undoubtedly have taken on a very different character regardless of the local environment. Hence, the "Mesopotamian advantage" was not a natural one but a sociopolitically engineered one.

A second decision that was made in bringing theories of landscape to bear on early complex polities was to accord a central analytical place to political authority—terrain traditionally occupied by the various economic determinants that have long held a privileged epistemological status within archaeology. In arguing against other forms of reductionism—ecological, economic—would it be fair to say that this work forwards an all-embracing vision of politics? It is certainly true that this book argues for a certain generality to political authority's reliance on landscape and that this can be read as a form of universalist argument. However, in contrast to absolutist visions that root the universal in an abiding faith in metahistory and subjectivist accounts that must presume a universality of the mind, the relational position forwarded here roots its general claims solely in a theory of political practice. It makes no claims on how landscape and politics intersect in any given case, merely that they do. Furthermore, politics here does not float unmoored from specific social locations but is instead located within a specific sociology of political relationships. This perspective is intended to allow for general theory that does not damage particularity. Indeed, the site of analysis is quite firmly located in the very nuances of each case because this is where the work of politics is ultimately done. Politics is not directly about territory, or urbanism, or architecture. It is about the production and reproduction of authority. However, territory, urbanism, and architecture *are* about politics; authority is profoundly constituted in the ordering of landscapes.

A third decision that was made in outlining the theoretical terrain of the present work centers on the position of resistance within the political landscape. Although a great deal of effort has gone into providing a cartography of political authority, I have given far less attention to the points where we might locate resistance. I have adopted this perspective quite intentionally. In the opening to his remarkable study of the transformation of the American advertising industry in the 1960s and 1970s, Thomas Frank justified his focus on advertising agencies rather than the readers of advertisements in the following terms: *"The Conquest of Cool* is a study of cultural production rather than reception, of power rather than resistance. . . . While cultural reception is a fascinating subject I hope that the reader will forgive me for leaving it to others. Not only has it been overdone, but our concentration on it, it seems to me, has led us to overlook and even minimize the equally fascinating doings of the creators of mass culture" (1997: x). This book has been more directly concerned with the constitution of political authority than with its corrosion and contestation, but this does not leave us with a totalizing vision of politics. Indeed, only by locating politics in space can we adequately plot a cartographic vision of challenge and resistance. Only once this element of the political landscape has been charted can we begin to discern the gaps in authority and the sites of challenge. In other words, how can we understand the sites of recent protests against the World Trade Organization without first understanding the geography of globalization?

It is one of the more intriguing peculiarities of recent intellectual history that modernism, a movement steadfastly determined to rid civil life of the accumulated burdens of the past, also shepherded the development of an intensive study of the ancient roots of human political associations that has revived the memory of rulers, institutions, and polities long forgotten. What possible import could, for example, the institutional organization of the Assyrian empire or the ritual duties of Maya kings hold for the modern who declares with Stephen Daedalus in James Joyce's *Ulysses,* "[H]istory is a nightmare from which I am trying to awake"? The answers to this question are undoubtedly numerous and multifaceted, but the present work has attempted to provide one perspective on how ancient political landscapes might provide a challenge to the modern civil order, even if it is only to make the political in the landscape slightly less dim in our minds and more vivid in our actions.

References Cited

Abrams, P. 1988. Notes on the Difficulty of Studying the State (1977). *Journal of Historical Sociology* 1(1):58–89.

Adams, R. E. W. 1986. Rio Azul. *National Geographic* 169:420–51.

Adams, R. McC. 1960. Early Civilizations, Subsistence, and Environment. In *City Invincible: A Symposium on Urbanization and Cultural Development in the Ancient Near East,* edited by C. H. Kraeling and R. M. Adams, 269–95. Chicago: University of Chicago Press.

——. 1965. *Land Behind Baghdad: A History of Settlement on the Diyala Plains.* Chicago: University of Chicago Press.

——. 1966. *The Evolution of Urban Society.* New York: Aldine.

——. 1978. Strategies of Maximization, Stability and Resilience in Mesopotamian Society. *Proceedings of the American Philosophical Society* 122(5):329–35.

——. 2000. Scale and Complexity in Archaic States. *Latin American Antiquity* 11:187–93.

——. n.d. Imperial Irrigation on the Eastern Mesopotamian Plain: The Kaskar Region in the Late Sasanian Period. Paper presented in the February 20, 2002, meeting of the University of Chicago Mesopotamian Irrigation Systems Seminar.

Adams, R. M., and H. G. Nissen. 1972. *The Uruk Countryside.* Chicago: University of Chicago Press.

Adontz, N. 1946. *Histoire d'Armenie.* Paris: L'Union General Armenienne de Bienfaisance.

Adorno, T. W. 1973. *The Jargon of Authenticity.* Evanston, Ill.: Northwestern University Press.

——. 1977. Letters to Walter Benjamin. In *Aesthetics and Politics,* edited by E. Bloch, G. Lukacs, B. Brecht, W. Benjamin, and T. Adorno, 110–33. London: Verso.

Adzhan, A. A., L. T. Gyuzalian, and B. B. Piotrovskii. 1932. Tsiklopichesckii Kreposti Zakavkaz'ya. *Soobshchenia Gosudarstvennoi Akademii Istorii Material'noi Kultury* 1(2):61–64.

Agnew, J. A. 1999. The New Geopolitics of Power. In *Human Geography Today,* edited by D. Massey, J. Allen, and P. Sarre, 173–93. Cambridge, Engl.: Polity.

Agnew, J. A., and S. Corbridge. 1995. *Mastering Space: Hegemony, Territory and International Political Economy.* London: Routledge.

Alberti, L. B. 1988. *On the Art of Building in Ten Books.* Translated by Joseph Rykwert, Neil Leach, and Robert Tavernor. Cambridge, Mass.: MIT Press.

Aldred, C. 1984. *The Egyptians.* London: Thames and Hudson.

Alexander, H. G., ed. 1956. *The Leibniz-Clarke Correspondence, Together with Extracts from Newton's Principia and Opticks.* Manchester: Manchester University Press.

Algaze, G. 1993. *The Uruk World System.* Chicago: University of Chicago Press.

———. 2001a. The Prehistory of Imperialism: The Case of Uruk Period Mesopotamia. In *Uruk Mesopotamia and Its Neighbors: Cross-Cultural Interactions in the Era of State Formation,* edited by M. S. Rothman, 27–83. Santa Fe: School of American Research Press.

———. 2001b. Initial Social Complexity in Southwestern Asia: The Mesopotamian Advantage. *Current Anthropology* 42(2):199–233.

Alonso, A. 1994. The Politics of Space, Time and Substance: State Formation, Nationalism, and Identity. *Annual Review of Anthropology* 24:379–405.

Altheim, F., and R. Altheim-Stiehl. 1954. *Ein Asiatischer Staat: Feudalismus unter den Sasaniden und ihren Nachbarn.* Wiesbaden: Limes-Verlag.

Althusser, L. 1971. *Lenin and Philosophy, and Other Essays,* translated by B. Brewster. New York: Monthly Review Press.

Anderson, B. 1983. *Imagined Communities.* London: Verso.

Andrae, W. 1938. *Das Wiedererstandene Assur.* Leipzig: J. C. Hinrichs.

Ankersmit, F. R. 1996. *Aesthetic Politics: Political Philosophy Beyond Fact and Value.* Stanford: Stanford University Press.

Appadurai, A. 1996. *Modernity at Large.* Minneapolis: University of Minnesota Press.

Arendt, H. 1951. *The Origins of Totalitarianism.* New York: Harcourt Brace.

———. 1958. What Was Authority? In *Nomos I: Authority,* edited by C. Friedrich, 81–112. Cambridge, Mass.: Harvard University Press.

Areshian, G., K. Kafadarian, A. Simonian, G. Tiratsian, and A. Kalantarian. 1977. Arkheologicheskie Issledovaniya v Ashtarakskom i Nairiskom Raionakh Armyanskoi SSR. *Vestnik Obshchesvennikh Nauk* 4:77–93.

Aristotle. 1988. *The Politics.* London: Penguin Books.

Ashmore, W. 1986. Peten Cosmology in the Maya Southeast: An Analysis of Architecture and Settlement Patterns at Classic Quirigua. In *The Southeast Maya Periphery,* edited by P. A. Urban and E. M. Schortman, 35–49. Austin: University of Texas Press.

———. 1989. Construction and Cosmology: Politics and Ideology in Lowland Maya Settlement Patterns. In *Word and Image in Maya Culture: Explorations in Language, Writing, and Representation,* edited by W. F. Hanks and D. S. Rice, 272–86. Salt Lake City: University of Utah Press.

———. 1991. Site-Planning Principles and Concepts of Directionality Among the Ancient Maya. *Latin American Antiquity* 2:199–226.

———. 1996. Authority and Assertion: Ancient Maya Politics and the Upper Belize Valley. Paper presented at the 95th Annual Meeting of the American Anthropological Association, San Francisco.

Auerbach, F. 1913. Das Gesertz der Bevölkerungskontration. *Petermanns Geographische Mitteilungen* 59: 74–76.

Avetisyan, H. 2001. *Aragats: Excavations of the Urartian Fortress.* Yerevan, Armenia: Yerevan State University Publications.

Avetisyan, P., R. Badalyan, and A. T. Smith. 2000. Preliminary Report on the 1998 Archaeological Investigations of Project ArAGATS in the Tsakahovit Plain, Armenia. *Studi Micenei ed Egeo-Anatolici* 42(1):19–59.

Azarpay, G. 1968. *Urartian Art and Artifacts: A Chronological Study.* Berkeley: University of California Press.

Bachelard, G. 1969. *The Poetics of Space.* Boston: Beacon Press.

Badaljan, R., C. Edens, P. Kohl, and A. Tonikijan. 1992. Archaeological Investigations at Horom in the Shirak Plain of Northwestern Armenia. *Iran* 30:31–48.

Badaljan, R., C. Edens, R. Gorny, P. L. Kohl, D. Stronach, A. V. Tonikajan, S. Hamayakjan, S. Mandrikjan, and M. Zardarjan. 1993. Preliminary Report on the 1992 Excavations at Horom, Armenia. *Iran* 31:1–24.

Badaljan, R. S., P. L. Kohl, and S. E. Kroll. 1997. Horom 1995. *Archaologische Mitteilungen Aus Iran und Turan* 29:191–228.

Badalyan, R., A. T. Smith, and P. Avetisyan. 2003. The Emergence of Socio-Political Complexity in Southern Caucasia. In *Archaeology in the Borderlands: Investigations in Caucasia and Beyond,* edited by A. T. Smith and K. Rubinson. Los Angeles: Cotsen Institute of Archaeology at UCLA.

Baines, J. 1995. Kingship, Definition of Culture, Legitimation. In *Ancient Egyptian Kingship,* edited by D. O'Connor and D. Silverman, 3–47. Leiden: Brill.

Baines, J., and N. Yoffee. 1998. Order, Legitimacy, and Wealth in Ancient Egypt and Mesopotamia. In *Archaic States,* edited by G. Feinman and J. Marcus, 199–260. Santa Fe: School of American Research Press.

———. 2000. Order, Legitimacy, and Wealth: Setting the Terms. In *Order, Legitimacy, and Wealth in Ancient States,* edited by J. E. Richards and M. Van Buren, 13–17. Cambridge, Engl.: Cambridge University Press.

Balakian, P. 1997. *Black Dog of Fate: A Memoir.* New York: Basic Books.

Ball, J. W., and J. T. Taschek. 1991. Late Classic Lowland Maya Political Organization and Central-Place Analysis. *Ancient Mesoamerica* 2:149–65.

Bard, K. 1992. Toward an Interpretation of the Role of Ideology in the Evolution of Complex Society in Egypt. *Journal of Anthropological Archaeology* 11:1–24.

Barkan, L. 1999. *Unearthing the Past: Archaeology and Aesthetics in the Making of Renaissance Culture.* New Haven: Yale University Press.

Barnes, G. L. 1999. *The Rise of Civilization in East Asia.* London: Thames and Hudson.

Barnett, R. D. 1950. The Excavations of the British Museum at Toprak Kale Near Van. *Iraq* 12(1):1–43.

Basso, K. H. 1996. *Wisdom Sits in Places: Landscape and Language Among the Western Apache*. Albuquerque: University of New Mexico Press.

Bataille, G. 1929. Dictionnaire Critique—Architecture. *Documents* 1:117.

Bawden, G. 1989. The Andean State as a State of Mind. *Journal of Anthropological Research* 45:327–32.

Beale, H. K. 1956. *Theodore Roosevelt and the Rise of America to World Power.* Baltimore: Johns Hopkins Press.

Bell, D. 1975. *Power, Influence, and Authority: An Essay in Political Linguistics.* New York: Oxford University Press.

Belli, O. 1980. Urartular'da Hayat Ağaci İnancı. *Anadolu Araştırmaları* 8:237–47.

Bender, B. 1998. *Stonehenge: Making Space.* Oxford: Berg.

Bendix, J., T. Mitchell, B. Ollman, and B. Sparrow. 1992. Going Beyond the State? *American Political Science Review* 86(4):1007–21.

Benjamin, W. 1977. *The Origin of German Tragic Drama.* Translated by J. Osborne. London: NLB.

———. 1985. Theses on the Philosophy of History. In *Illuminations,* edited by H. Arendt. New York: Schocken Books.

Berlin, H. 1958. El Glifo "emblema" en las Inscripziones Mayas. *Journal de la Société des Américanistes* 47:111–19.

Berlin, I. 1976. *Vico and Herder: Two Studies in the History of Ideas.* New York: Viking Press.

Bermingham, A. 1986. *Landscape and Ideology: The English Rustic Tradition, 1740–1860.* Berkeley: University of California Press.

Berry, B. J. L., and A. Pred. 1965. *Central Place Studies: A Bibliography of Theory and Applications.* Philadelphia: Regional Science Research Institute.

Bilgiç, E. and B. Öğün. 1967. Excavations at Kef Kalesi of Adilcevaz, 1964. *Anatolia (Anadolu)* 9:11–19.

Binford, L. R. 1965. Archaeological Systematics and the Study of Cultural Process. *American Antiquity* 31:203–10.

Bintliff, J. L., ed. 1991. *The Annales School and Archaeology.* Leicester: Leicester University Press.

Biscione, R. 1994. Missione Archeologica Italo-Armena nel Territorio del Lago Sevan Campagna 1994. *Studi Micenei ed Egeo-Anatolici* 34:146–49.

———. Forthcoming. Pre-Urartian and Urartian Settlement Patterns in the Caucasus, Two Case Studies: The Urmia Plain, Iran, and the Southern Sevan Basin, Armenia. In *Archaeology in the Borderlands: Investigations in Caucasia and Beyond,* edited by A. T. Smith and K. Rubinson. Los Angeles: Cotsen Institute of Archaeology at UCLA.

Blanton, R. E. 1976. Anthropological Studies of Cities. *Annual Review of Anthropology* 5:249–64.

Blanton, R., and G. Feinman. 1984. The Mesoamerican World System. *American Anthropologist* 86:673–82.

Blau, P. M. 1977. *Inequality and Heterogeneity: A Primitive Theory of Social Structure.* New York: Free Press.

Blier, S. P. 1987. *The Anatomy of Architecture: Ontology and Metaphor in Batamaliba Architectural Expression.* Cambridge, Engl.: Cambridge University Press.

Bloch, M. 1987. The Ritual of the Royal Bath in Madagascar: The Dissolution of Death, Birth and Fertility into Authority. In *Rituals of Royalty: Power and Ceremonial in Traditional Societies,* edited by D. Cannadine and S. Price, 271–97. Cambridge, Engl.: Cambridge University Press.

Bloomer, K., and C. Moore. 1977. *Body, Memory, and Architecture.* New Haven: Yale University Press.

Boone, E. H. 1998. Maps of Territory, History, and Community in Aztec Mexico. In *Cartographic Encounters,* edited by G. M. Lewis, 111–33. Chicago: University of Chicago Press.

Bossuet, J. B. 1990. *Politics Drawn from the Very Words of Holy Scripture.* Cambridge, Engl.: Cambridge University Press.

Bourdieu, P. 1977. *Outline of a Theory of Practice.* Cambridge, Engl.: Cambridge University Press.

———. 1999. Rethinking the State: Genesis and Structure of the Bureaucratic Field. In *State/Culture: State Formation After the Cultural Turn,* edited by G. Steinmetz, 53–75. Ithaca, N.Y.: Cornell University Press.

Bradley, R. 1998. *The Significance of Monuments: On the Shaping of Human Experience in Neolithic and Bronze Age Europe.* London: Routledge.

———. 2000. *An Archaeology of Natural Places.* London: Routledge.

Brady, J. E., and W. Ashmore. 1999. Mountains, Caves, Water: Ideational Landscapes of the Ancient Maya. In *Archaeologies of Landscape,* edited by W. Ashmore and A. B. Knapp, 124–45. Oxford: Blackwell.

Braidwood, R. J. 1964. *Prehistoric Men.* Glenview, Ill.: Scott Foresmen.

Brand, S. 1994. *How Buildings Learn: What Happens After They're Built.* New York: Viking.

Braudel, F. 1972–73. *The Mediterranean and the Mediterranean World in the Age of Philip II.* Translated by S. Reynolds. New York: Harper and Row.

Breasted, J. H. 1919. The Origins of Civilization. *The Scientific Monthly* 10:268–89.

Buck-Morss, S. 1989. *The Dialectics of Seeing: Walter Benjamin and the Arcades Project.* Cambridge, Mass.: MIT Press.

Bullard, W. R. 1960. Maya Settlement Pattern in Northeastern Peten, Guatemala. *American Antiquity* 25:355–72.

Burka, P. 1994. George W. Bush and the New Political Landscape: How the Republicans Took Over Texas—and What It Means. *Texas Monthly* 22(12):128–29.

Burney, C. 1993. The God Haldi and the Urartian State. In *Aspects of Art and Iconography: Anatolia and Its Neighbors: Studies in Honor of Nimet Özgüç,* edited by M. J. Mellink, E. Porada, and T. Özgüç, 107–10. Ankara, Turkey: Türk Tarih Kurumu Basımevi.

Butzer, K. 1980. Civilizations: Organisms or Systems? *American Scientist* 68:517–23.

Calmeyer, P. 1991. Some Remarks on Iconography. In *Urartu: A Metalworking Center in the First Millennium B.C.E.,* edited by R. Merhav, 311–19. Jerusalem: The Israel Museum.

Canby, J. V. 2001. *The Ur-Nammu Stela.* Philadelphia: University of Pennsylvania Museum of Archaeology and Anthropology.

Cannon, S. F. 1978. *Science in Culture: The Early Victorian Period.* New York: Science History Publications.

Carneiro, R. 1970. A Theory of the Origin of the State. *Science* 169:733–39.

———. 1981. The Chiefdom as Precursor of the State. In *The Transition to State-hood in the New World,* edited by G. D. Jones and R. Kautz, 37–79. Cambridge, Engl.: Cambridge University Press.

Casey, E. S. 1997. *The Fate of Place.* Berkeley: University of California Press.

Castagnoli, F. 1971. *Orthogonal Town Planning in Antiquity.* Cambridge, Mass.: MIT Press.

Castells, M. 1983. *The City and the Grassroots.* Berkeley: University of California Press.

Champion, T. C., ed. 1989. *Centre and Periphery in the Ancient World.* London: Unwin Hyman.

Chapman, J. 1997. Landscapes in Flux and the Colonization of Time. In *Landscapes in Flux: Central and Eastern Europe in Antiquity,* edited by J. Chapman and P. Dolukhanov, 1–22. Oxford: Oxbow Books.

Chase, A. F., and D. Z. Chase. 1996. More Than Kin and King: Centralized Political Organization Among the Late Classic Maya. *Current Anthropology* 37(5):803–10.

Cherry, J. F. 1987. Power in Space: Archaeological and Geographical Studies of the State. In *Landscape and Culture,* edited by J. M. Wagstaff, 146–72. Oxford: Basil Blackwell.

Cherry, J. F., J. L. Davis, and E. Mantzourani. 1991. *Landscape Archaeology as Long-Term History: Northern Keos in the Cycladic Islands.* Los Angeles: UCLA Institute of Archaeology.

Childe, V. G. 1931. *Skara Brae, a Pictish Village in Orkney.* London: Kegan Paul.

———. 1936. *Man Makes Himself.* London: Watts.

———. 1946. *What Happened in History.* New York: Penguin Books.

———. 1950. The Urban Revolution. *Town Planning Review* 21:3–17.

Christaller, W., and C. W. Baskin. 1966. *Central Places in Southern Germany.* Englewood Cliffs, N.J.: Prentice Hall.

Çilingiroğlu, A. 1983. Mass Deportations in the Urartian Kingdom. *Anadolu Araştırmaları* 9:319–23.

Civil, M. 1987. Ur III Bureaucracy: Quantitative Aspects. In *The Organization of Power: Aspects of Bureaucracy in the Ancient Near East,* edited by M. Gibson and R. D. Biggs, 43–53. Chicago: Oriental Institute of the University of Chicago.

Claessen, H. J. M. 1978. The Early State: A Structural Approach. In *The Early State,* edited by H. J. M. Claessen and P. Skalnik, 533–96. The Hague: Mouton.

———. 1984. The Internal Dynamics of the Early State. *Current Anthropology* 25(4):365–79.

Claessen, H. J. M., and P. Skalnik. 1978. The Early State: Theories and Hypotheses. In *The Early State,* edited by H. J. M. Claessen and P. Skalnik, 3–29. The Hague: Mouton.

Claessen, H. J. M., P. v. d. Velde, and M. E. Smith. 1985. *Development and Decline: The Evolution of Sociopolitical Organizations.* South Hadley, Mass.: Bergin and Garvey.

Clark, S. 1985. The *Annales* Historians. In *The Return of Grand Theory in the Human Sciences*, edited by Q. Skinner, 177–98. Cambridge, Engl.: Cambridge University Press.

Clark, T. 1991. *Charles Olson: The Allegory of a Poet's Life*. New York: W. W. Norton.

Clarke, D. L. 1968. *Analytical Archaeology*. London: Methuen.

Clegg, S. 1989. *Frameworks of Power*. London: Sage.

Coe, M. D. 1999. *The Maya*. London: Thames and Hudson.

Coggins, C. 1980. The Shape of Time: Some Political Implications of a Four-Part Figure. *American Antiquity* 45:727–39.

Cohen, Roger. 2001. New Chancellery in Berlin Is a Colossus. What Would Albert Speer Say? In *The New York Times*, Feb. 16.

Cohen, Ronald. 1978. State Origins: A Reappraisal. In *The Early State*, edited by H. J. M. Claessen and P. Skalnik, 31–75. The Hague: Mouton.

Cohen, Ronald, and E. R. Service, eds. 1978. *Origins of the State*. Philadelphia: Institute for the Study of Human Issues.

Collingwood, R. G. 1994. *The Idea of History*. Oxford: Oxford University Press.

Comaroff, J., and J. L. Comaroff. 1997. *Of Revelation and Revolution, Volume II: The Dialectics of Modernity on a South African Frontier*. Chicago: University of Chicago Press.

Comaroff, J. L. 1998. Reflections on the Colonial State in South Africa and Elsewhere: Factions, Fragments, Facts and Fictions. *Social Identities* 4(3):321–61.

Conrad, G. W., and A. A. Demarest. 1984. *Religion and Empire: The Dynamics of Aztec and Inca Expansionism*. Cambridge, Engl.: Cambridge University Press.

Cooper, J. S. 1986. *Presargonic Inscriptions*. Sumerian and Akkadian royal inscriptions; vol. 1. New Haven, Conn.: American Oriental Society.

Corrigan, P., and D. Sayer. 1985. *The Great Arch*. Oxford: Basil Blackwell.

Cosgrove, D. E. 1993. *The Palladian Landscape: Geographical Change and Its Cultural Representations in Sixteenth-Century Italy*. Leicester: Leicester University Press.

Cosgrove, D. E., and S. Daniels. 1988. *The Iconography of Landscape: Essays on the Symbolic Representation, Design, and Use of Past Environments*. Cambridge, Engl.: Cambridge University Press.

Cribb, R. 1991. *Nomads in Archaeology*. Cambridge, Engl.: Cambridge University Press.

Crumley, C. L., and W. H. Marquardt. 1990. Landscape: A Unifying Concept in Regional Analysis. In *Interpreting Space: GIS and Archaeology*, edited by K. M. S. Allen, S. W. Green, and E. B. W. Zubrow, 73–79. London: Taylor and Francis.

Culbert, T. P. 1988. The Collapse of Classic Maya Civilization. In *The Collapse of Ancient States and Civilizations*, edited by N. Yoffee and G. L. Cowgill, 69–101. Tucson: University of Arizona Press.

———. 1991. Maya Political History and Elite Interaction: A Summary View. In *Classic May Political History*, edited by T. P. Culbert. Albuquerque: University of New Mexico Press.

Dahl, R. A. 1961. *Who Governs? Democracy and Power in an American City*. New Haven: Yale University Press.

Dahrendorf, R. 1959. *Class and Class Conflict in Industrial Society.* Stanford: Stanford University Press.

Darnton, R. 1984. *The Great Cat Massacre and Other Episodes in French Cultural History.* New York: Basic Books.

Davis, K. 1965. The Urbanization of the Human Population. *Scientific American* 213(3):40–53.

Davis, M. 1990. *City of Quartz: Excavating the Future in Los Angeles.* New York: Verso.

——. 1998. *Ecology of Fear: Los Angeles and the Imagination of Disaster.* New York: Metropolitan Books.

de Certeau, M. 1984. *The Practice of Everyday Life.* Berkeley: University of California Press.

de Coulanges, F. 1874. *The Ancient City,* translated by W. Small. Boston: Lee and Shepard.

Deimel, A. 1931. *Sumerische Tempelwirtschaft zur Zeit Urukaginas und Seiner Vorgänger.* Rome: Pontifleio Istituto Biblico.

Demarest, A. A. 1992. Ideology in Ancient Maya Cultural Evolution: The Dynamics of Galactic Polities. In *Ideology and the Evolution of Precolumbian Civilizations,* edited by A. Demarest and G. Conrad. Cambridge, Engl.: Cambridge University Press.

de Montmollin, O. 1989. *The Archaeology of Political Structure.* Cambridge, Engl.: Cambridge University Press.

Dever, W. 1997. Biblical Archaeology. In *The Oxford Encyclopedia of Archaeology in the Near East,* edited by E. M. Meyers, 315–19. Oxford: Oxford University Press.

Dewey, J. 1927. *The Public and Its Problems.* London: George Allen and Unwin.

Diakonoff, I. M. 1969. The Rise of the Despotic State in Ancient Mesopotamia. In *Ancient Mesopotamia, Socio-Economic History,* edited by I. M. Diakonoff, 173–203. Moscow: Nauka.

——. 1991. General Outline of the First Period of the History of the Ancient World and the Problem of the Ways of Development. In *Early Antiquity,* edited by I. M. Diakonoff, 27–66. Chicago: University of Chicago Press.

Diamond, S. 1974. *In Search of the Primitive: A Critique of Civilization.* New Brunswick, N.J.: Transaction Books.

Dietler, M. 1998. Consumption, Agency, and Cultural Entanglement: Theoretical Implications of a Mediterranean Colonial Encounter. In *Studies in Culture Contact: Interaction, Culture Change, and Archaeology,* edited by J. G. Cusick, 288–315. Carbondale: Southern Illinois University Press.

Dobres, M.-A., and J. E. Robb, eds. 2000. *Agency in Archaeology.* New York: Routledge.

Dowdall, H. 1923. The Word 'State.' *The Law Quarterly Review* 39:98–125.

Duara, P. 1996. Historicizing National Identity, or Who Imagines What and When. In *Becoming National: A Reader,* edited by G. Eley and R. G. Suny. New York: Oxford University Press.

Dumond, D. E. 1972. Population Growth and Political Centralization. In *Population Growth: Anthropological Implications,* edited by B. Spooner, 286–310. Cambridge, Mass.: M.I.T. Press.

Duncan, J. S. 1990. *The City as Text: The Politics of Landscape Interpretation in the Kandyan Kingdom.* Cambridge, Engl.: Cambridge University Press.

Durkheim, E. 1986. *Durkheim on Politics and the State.* Stanford: Stanford University Press.

Eagleton, T. 1990. *The Ideology of the Aesthetic.* Oxford: Blackwell.

Earle, T. K. 1987. Specialization and the Production of Wealth: Hawaiian Chiefdoms and the Inka Empire. In *Specialization, Exchange, and Complex Societies,* edited by E. M. Brumfiel and T. K. Earle, 64–75. Cambridge, Engl.: Cambridge University Press.

——, ed. 1991. *Chiefdoms: Power, Economy, and Ideology.* Cambridge, Engl.: Cambridge University Press.

Earle, T. K., T. N. D'Altroy, C. J. LeBlanc, C. A. Hastorf, and T. Y. LeVine. 1980. Changing Settlement Patterns in the Upper Montaro Valley, Peru. *Journal of New World Archaeology* 4(1):1–49.

Easton, D. 1953. *The Political System, an Inquiry into the State of Political Science.* New York: Knopf.

——. 1968. Political Science. In *International Encyclopedia of the Social Sciences,* edited by D. L. Sills. New York: Macmillan.

Eck, C. van. 1994. *Organicism in Nineteenth-Century Architecture: An Inquiry into Its Theoretical and Philosophical Background.* Amsterdam: Architectura and Natura Press.

Eckhardt, W. 1995. A Dialectical Evolutionary Theory of Civilizations, Empires, and Wars. In *Civilizations and World Systems,* edited by S. K. Sanderson, 75–94. Walnut Creek, Calif.: Altamira Press.

Eisenman, P. 1984. The End of the Classical: The End of the Beginning, the End of the End. *Perspecta* 21:153–73.

Elkin, S. L. 1987. *City and Regime in the American Republic.* Chicago: University of Chicago Press.

Emberling, G. 1995. Ethnicity and the State in Early Third Millennium Mesopotamia. Ph.D. dissertation, University of Michigan.

Emery, D. B. 1986. Popular Music of the Clash: A Radical Challenge to Authority. In *The Frailty of Authority,* edited by M. J. Aronoff, 147–65. New Brunswick, N.J.: Transaction Books.

Engels, F. 1990. The Origin of Family, Private Property, and the State. In *Karl Marx–Frederick Engels: Collected Works,* vol. 26, translated by R. Dixon, 129–276. New York: International Publishers.

Erzen, A. 1988. *Çavuştepe I: Urartian Architectural Monuments of the 7th and 6th Centuries B.C.* and a Necropolis of the Middle Age. Ankara, Turkey: Türk Tarih Kurumu Basımevi.

Evans, P. B., D. Rueschemeyer, and T. Skocpol. 1985. On the Road Toward a More Adequate Understanding of the State. In *Bringing the State Back In,* edited by P. B. Evans, D. Rueschemeyer, and T. Skocpol, 347–66. Cambridge, Engl.: Cambridge University Press.

Falconer, S. E., and S. H. Savage. 1995. Heartlands and Hinterlands: Trajectories of Early Urbanization in Mesopotamia and the Southern Levant. *American Antiquity* 60(1):37–58.

Febvre, L. 1925. *A Geographical Introduction to History.* New York: Knopf.

Feldman, A. 1991. *Formations of Violence.* Chicago: University of Chicago Press.

Ferguson, T. J. 1996. *Historic Zuni Architecture and Society: An Archaeological Application of Space Syntax.* Tucson: University of Arizona Press.

Ferguson, Y. H., and R. W. Mansbach. 1996. *Polities: Authority, Identities, and Change.* Columbia: University of South Carolina Press.

Finkelstein, J. J. 1979. Early Mesopotamia, 2500–1000 B.C. In *Propaganda and Communication in World History. Volume 1: The Symbolic Instrument in Early Times,* edited by H. D. Lasswell, D. Lerner, and H. Spier, 60–63. Honolulu: University Press of Hawaii.

Finley, M. I. 1977. The Ancient City: From Fustel de Coulanges to Max Weber and Beyond. *Comparative Studies in Society and History* 19:305–27.

———. 1981. Politics. In *The Legacy of Greece: A New Appraisal,* edited by M. I. Finley and R. W. Livingstone, 22–36. Oxford: Clarendon Press.

Flannery, K. V. 1972. The Cultural Evolution of Civilizations. *Annual Review of Ecology and Systematics* 3:399–426.

———. 1977. Review of *Mesoamerican Archaeology: New Approaches. American Antiquity* 42(4):659–61.

———. 1998. The Ground Plans of Archaic States. In *Archaic States,* edited by G. M. Feinman and J. Marcus, 15–57. Santa Fe: School of American Research.

Folan, W. J., J. Marcus, and W. F. Miller. 1995. Verification of a Maya Settlement Model Through Remote Sensing. *Cambridge Archaeological Journal* 5(2):277–83.

Forbes, T. 1983. *Urartian Architecture.* Oxford: BAR International.

Fortes, M., and E. E. Evans-Pritchard. 1940. *African Political Systems.* London: Oxford University Press.

Foster, B. 1981. A New Look at the Sumerian Temple State. *Journal of the Economic and Social History of the Orient* 24:225–41.

———. 1996. *Before the Muses: An Anthology of Akkadian Literature.* Bethesda, Md.: CDL Press.

Foster, H. 1985. *Recodings: Art, Spectacle, Cultural Politics.* Port Townsend, Wash.: Bay Press.

Foucault, M. 1978. *The History of Sexuality,* vol. 1. London: Penguin Books.

———. 1979a. *Discipline and Punish: The Birth of the Prison.* New York: Vintage Books.

———. 1979b. On Governmentality. *Ideology and Consciousness* 6:5–21.

———. 1982. The Subject and Power. In *Michel Foucault: Beyond Structuralism and Hermeneutics,* edited by H. Dreyfus and P. Rabinow, 208–26. Chicago: University of Chicago Press.

———. 1984. Space, Knowledge, and Power. In *The Foucault Reader,* edited by P. Rabinow, 239–56. New York: Penguin Books.

Fox, J. W., G. W. Cook, A. F. Chase, and D. Z. Chase. 1996. Questions of Political and Economic Integration: Segmentary Versus Centralized States Among the Ancient Maya. *Current Anthropology* 37(5):795–801.

Frampton, K. 1992. *Modern Architecture: A Critical History.* London: Thames and Hudson.

Frangipane, M. 2001. Centralization Processes in Greater Mesopotamia: Uruk "Expansion" as the Climax of Systemic Interactions Among Areas of the Greater Mesopotamian Region. In *Uruk Mesopotamia and Its Neighbors: Cross-Cultural Interactions in the Era of State Formation,* edited by M. S. Rothman, 307–47. Santa Fe: School of American Research Press.

Frank, A. G. 1993. Bronze Age World System Cycles. *Current Anthropology* 34(4): 383–429.

Frank, T. 1997. *The Conquest of Cool.* Chicago: University of Chicago Press.

———. 2000. *One Market Under God: Extreme Capitalism, Market Populism, and the End of Economic Democracy.* New York: Doubleday.

Frankfort, H. 1996. *The Art and Architecture of the Ancient Orient.* New Haven: Yale University Press.

Frayne, D. 1997. *Ur III period (2112–2004 B.C.).* The Royal Inscriptions of Mesopotamia: Early Periods vol. 3/2. Toronto: University of Toronto Press.

Freud, S. 1961. *The Future of an Illusion.* New York: W. W. Norton.

Fried, M. H. 1967. *The Evolution of Political Society: An Essay in Political Anthropology.* New York: Random House.

Friedrich, C. J. 1958. Authority, Reason, and Discretion. In *Nomos I: Authority,* edited by C. J. Friedrich, 28–48. Cambridge, Mass.: Harvard University Press.

Gadamer, H.-G. 1975. *Truth and Method.* New York: Crossroad.

Gallagher, W. R. 1994. Assyrian Deportation Propaganda. *State Archives of Assyria Bulletin* 7(2):57–65.

Garber, D. 1995. Leibniz: Physics and Philosophy. In *The Cambridge Companion to Leibniz,* edited by N. Jolley, 270–352. Cambridge, Engl.: Cambridge University Press.

Geertz, C. 1980. *Negara: The Theater-State in Nineteenth Century Bali.* Princeton: Princeton University Press.

Gelb, I. J. 1969. On the Alleged Temple and State Economies in Ancient Mesopotamia. In *Studi in Onore di Edoardo Volterra,* vol. 6, 137–54. Milan: A Giuffré.

Gellner, E. 1994. *Encounters with Nationalism.* Oxford: Blackwell.

George, A. 1999. *The Epic of Gilgamesh.* London: Allen Lane.

Gibson, M. 2000. Hamoukar: Early City in Northeastern Syria. *The Oriental Institute News and Notes* 166.

Giddens, A. 1976. *New Rules of Sociological Method: A Positive Critique of Interpretative Sociologies.* New York: Basic Books.

———. 1984. *The Constitution of Society: Outline of the Theory of Structuration.* Cambridge, Engl.: Polity.

———. 1985. *The Nation-State and Violence.* Cambridge, Engl.: Polity.

Giddens, A., M. Mann. and I. Wallerstein. 1989. Review Symposium: Comments on Paul Kennedy's The Rise and Fall of the Great Powers. *The British Journal of Sociology* 40(2):328–40.

Gillespie, S. D. 2000. Rethinking Ancient Maya Social Organization Replacing "Lineage" with "House." *American Anthropologist* 102(3):467–84.

Glacken, C. 1967. *Traces on the Rhodian Shore: Nature and Culture in Western*

Thought from Ancient Times to the End of the Eighteenth Century. Berkeley: University of California Press.

Glassie, H. 1975. *Folk Housing in Middle Virginia.* Knoxville: University of Tennessee Press.

Glassie, H. 1977. Archaeology and Folklore: Common Anxieties, Common Hopes. In *Historical Archaeology and the Importance of Material Things,* edited by L. Ferguson, 23–35. Lansing, Mich.: Society for Historical Archaeology.

Golden, C. W. 2003. The Politics of Warfare in the Usumacinta Basin: La Pasadita and the Realm of Bird Jaguar. In *Ancient Mesoamerican Warfare,* edited by T. Stanton and M. K. Brown. Walnut Creek, Calif.: Altamira Press.

Gombrich, E. H. 1966. *Norm and Form: Studies in the Art of the Renaissance.* London: Phaidon.

Goodsell, C. T. 1988. *The Social Meaning of Civic Space: Studying Political Authority Through Architecture.* Lawrence: University Press of Kansas.

Gordon, D. M. 1977. *Problems in Political Economy: An Urban Perspective.* Lexington, Mass.: Heath.

Gordon, D. M., R. Edwards, and M. Reich. 1982. *Segmented Work, Divided Workers: The Historical Transformations of Labor in the United States.* Cambridge, Engl.: Cambridge University Press.

Gottfried, P. 1995. Reconfiguring the Political Landscape. *Telos* 103:111–27.

Graham, I. 1996. *Corpus of Maya Hieroglyphic Inscriptions,* vol. 7, part 1. Cambridge, Mass.: Peabody Museum, Harvard University.

Gramsci, A. 1971. *Selections from the Prison Notebooks.* New York: International Publishers.

Gray, J. 1992. Against the New Liberalism. *Times Literary Supplement* 4657:13–15.

Grayson, A. K. 1987. *Assyrian Rulers of the Third and Second Millennia B.C. (to 1115 B.C.).* Toronto: University of Toronto Press.

Green, P. 1996. The Political Institutions of the Good Society. In *The Constitution of Good Societies,* edited by K. E. Soltan and S. L. Elkin, 164–85. University Park: Pennsylvania State University Press.

Grosby, S. 1995. Territoriality: The Transcendental Primordial Feature of Modern Societies. *Nations and Nationalism* 1(2):143–62.

———. 1997. Borders, Territory and Nationality in the Ancient Near East and Armenia. *Journal of the Economic and Social History of the Orient* 40(1):1–29.

Guillemin, G. F. 1968. Development and Function of the Tikal Ceremonial Center. *Ethnos* 33:1–35.

Gumerman, G. J., and M. Gell-Mann, eds. 1994. *Understanding Complexity in the Prehistoric Southwest.* Reading, Mass.: Addison-Wesley.

Haas, J. 1982. *The Evolution of the Prehistoric State.* New York: New York University Press.

Hall, E. T. 1966. *The Hidden Dimension.* Garden City, N.Y.: Doubleday.

Hallo, W. W., and W. K. Simpson. 1971. *The Ancient Near East.* San Diego, Calif.: Harcourt Brace Jovanovich.

Hammond, N. 1972. Locational Models and the Site of Lubaantun: A Classic Maya Centre. In *Models in Archaeology,* edited by D. L. Clarke, 757–800. London: Methuen.

——. 1974. The Distribution of Late Classic Maya Major Ceremonial Centers in the Central Area. In *Mesoamerican Archaeology: New Approaches,* edited by N. Hammond, 313–34. Austin: University of Texas Press.

Hannerz, U. 1992. *Cultural Complexity.* New York: Columbia University Press.

Harris, M. 1977. *Cannibals and Kings: The Origins of Cultures.* New York: Random House.

——. 1979. *Cultural Materialism.* New York: Random House.

Harris, R. 1975. *Ancient Sippar: A Demographic Study of an Old Babylonian City.* Leiden: Nederlands Historisch-Archaeologisch Instituut te Istanbul.

Harrison, P. D. 1981. Some Aspects of Preconquest Settlement in Southern Quintana Roo, Mexico. In *Lowland Maya Settlement Patterns,* edited by W. Ashmore, 259–86. Albuquerque: University of New Mexico Press.

Hartshorne, R. 1939. *The Nature of Geography: A Critical Survey of Current Thought in the Light of the Past.* Lancaster, Pa.: Association of American Geographers.

Harvey, D. 1973. *Social Justice and the City.* Baltimore: Johns Hopkins University Press.

——. 1985a. *Consciousness and the Urban Experience: Studies in the History and Theory of Capitalist Urbanization.* Baltimore: John Hopkins University Press.

——. 1985b. *The Urbanization of Capital: Studies in the History and Theory of Capitalist Urbanization.* Baltimore: John Hopkins University Press.

——. 1989. *The Condition of Postmodernity.* Oxford: Basil Blackwell.

——. 1996. *Justice, Nature, and the Geography of Difference.* New York: Blackwell.

——. 2000. Cosmopolitanism and the Banality of Geographic Evils. *Public Culture* 12(2):529–64.

Hass, K. A. 1998. *Carried to the Wall: American Memory and the Vietnam Veterans Memorial.* Berkeley: University of California Press.

Hattenhauer, D. 1984. The Rhetoric of Architecture. *Communication Quarterly* 32(1):71–77.

Hegel, G. W. F. 1988. *Introduction to the Philosophy of History.* Indianapolis: Hackett.

Herder, J. G. 1966. *Outlines of a Philosophy of the History of Man.* New York: Bergman.

Hexter, J. H. 1972. Fernand Braudel and the *monde Braudellien. Journal of Modern History* 44:480–539.

Hill, J. N., and J. Gunn, eds. 1977. *The Individual in Prehistory: Studies of Variability in Style in Prehistoric Technologies.* New York: Academic Press.

Hillier, B., and J. Hanson. 1984. *The Social Logic of Space.* Cambridge, Engl.: Cambridge University Press.

Hinsley, F. H. 1986. *Sovereignty.* Cambridge, Engl.: Cambridge University Press.

Hirt, A. 1801. *Die Baukunst Nach den Grundsätzen der Alten.* Berlin: Realschulbuchhandlung.

Hmayakian, S. G. 1990. *Gosudarstvenaya Religiya Vanskovo Tsarstva.* Yerevan: Izdatelst'stvo AN Armenij.

Hmayakyan, S. G., V. A. Igumnov, and H. H. Karagyozyan. 1996. An Urartian Cuneiform Inscription from Ojasar-Ilandagh, Nakhichevan. *Studi Micenei ed Egeo-Anatolici* 37:139–51.

Hobbes, T. 1998. *On the Citizen.* Cambridge, Engl.: Cambridge University Press.

Hobhouse, L. T. 1911. *Social Evolution and Political Theory.* New York: Columbia University Press.

Hobhouse, L. T., G. C. W. C. Wheeler, and M. Ginsberg. 1915. *The Material Culture and Social Institutions of the Simpler Peoples: An Essay in Correlation.* London: Chapman and Hall.

Hobsbawm, E. J. 1990. *Nations and Nationalism Since 1780: Programme, Myth, Reality.* Cambridge, Engl.: Cambridge University Press.

Hodder, I. 1972. Locational Models and Romano-British Settlement. In *Models in Archaeology,* edited by D. L. Clarke, 887–909. London: Methuen.

——. 1986. *Reading the Past.* Cambridge, Engl.: Cambridge University Press.

——. 1990. *The Domestication of Europe.* Oxford: Basil Blackwell.

Hodder, I., and M. Hassall. 1971. The Non-Random Spacing of Romano-British Walled Towns. *Man* 6:391–407.

Hodder, I., and C. Orton. 1976. *Spatial Analysis in Archaeology.* Cambridge, Engl.: Cambridge University Press.

Hoebel, E. A. 1954. *The Law of Primitive Man.* New York: Atheneum.

Hoffman, J. 1998. *Sovereignty.* Minneapolis: University of Minnesota Press.

Holland, T. A. 1997. Jericho. In *The Oxford Encyclopedia of Archaeology in the Near East,* edited by E. M. Meyers, 220–24. Oxford: Oxford University Press.

Horden, P., and N. Purcell. 2000. *The Corrupting Sea: A Study of Mediterranean History.* Oxford: Blackwell.

Horne, L. 1994. *Village Spaces.* Washington, D.C.: Smithsonian Institution Press.

Hoskins, W. G. 1977. *The Making of the English Landscape.* London: Hodder and Stoughton.

Houston, S. D. 1993. *Hieroglyphs and History at Dos Pilas: Dynastic Politics of the Classic Maya.* Austin: University of Texas Press.

Howard, J. S. 1998. Subjectivity and Space: Deleuze and Guattari's BwO in the New World Order. In *Deleuze & Guattari: New Mappings in Politics, Philosophy, and Culture,* edited by E. Kaufman and K. J. Heller, 112–26. Minneapolis: University of Minnesota Press.

Humboldt, A. von. 1847. *Cosmos: A Sketch of a Physical Description of the Universe.* London: Longmans.

Humphreys, S. C. 1978. *Anthropology and the Greeks.* London: Routledge and Kegan Paul.

Huntington, S. P. 1993. The Clash of Civilizations? *Foreign Affairs* 72(3):22–49.

——. 1996. *The Clash of Civilizations and the Remaking of World Order.* New York: Simon and Schuster.

Inomata, T., and K. Aoyama. 1996. Central-Place Analyses in the La Entrada Region, Honduras: Implications for Understanding the Classic Maya Political and Economic Systems. *Latin American Antiquity* 7(4):291–312.

Isard, W. 1956. *Location and Space-Economy: A General Theory Relating to Industrial Location, Market Areas, Land Use, Trade, and Urban Structure.* Cambridge, Mass.: Technology Press of M.I.T.

Jackson, J. B. 1970. The Public Landscape. In *Landscapes,* edited by E. H. Zube, 153–60. Amherst: University of Massachusetts Press.

———. 1979. The Order of a Landscape: Reason and Religion in Newtonian America. In *The Interpretation of Ordinary Landscapes,* edited by D. W. Meining, 153–63. Oxford: Oxford University Press.

———. 1984. *Discovering the Vernacular Landscape.* New Haven: Yale University Press.

Jacobs, J. 1969. *The Economy of Cities.* New York: Random House.

Jacobsen, T. 1939. *The Sumerian King List.* Chicago: University of Chicago Press.

———. 1970. Early Political Development in Mesopotamia. In *Toward the Image of Tammuz,* edited by W. Moran, 132–56. Cambridge, Mass.: Harvard University Press.

———. 1987. Pictures and Pictorial Language (The Burney Relief). In *Figurative Language in the Ancient Near East,* edited by M. Mindlin, M. J. Geller, and J. E. Wansbrough, 1–11. London: School of Oriental and African Studies.

Jencks, C. 1973. *Le Corbusier and the Tragic View of Architecture.* Cambridge, Mass.: Harvard University Press.

———. 1991. *The Language of Post-Modern Architecture.* New York: Rizzoli.

Jessop, B. 1990. *State Theory: Putting the Capitalist State in Its Place.* Cambridge, Engl.: Polity.

Joffe, A. H. 1998. Disembedded Capitals in Western Asian Perspective. *Comparative Studies in Society and History* 40(3):549–80.

Johnson, A. W., and T. Earle. 1987. *The Evolution of Human Societies: From Foraging Group to Agrarian States.* Stanford: Stanford University Press.

Johnson, G. A. 1972. A Test of the Utility of Central Place Theory in Archaeology. In *Man, Settlement, and Urbanism,* edited by P. Ucko, R. Tringham, and G. Dimbleby, 769–85. London: Duckworth.

———. 1973. Local Exchange and Early State Development in Southwestern Iran. *Museum of Anthropology, University of Michigan, Anthropological Papers* 51.

Johnston, R. J. 1982. *Geography and the State: An Essay in Political Geography.* London: Macmillan Press.

Kafadarian, K. 1984. *Arkitektura Goroda Argishtihinili.* Yerevan: Izdatel'stvo Armianskoi SSR.

Kahn, C. H. 1960. *Anaximander and the Origins of Greek Cosmology.* New York: Columbia University Press.

Kalantar, A. 1994. *Armenia: From the Stone Age to the Middle Ages.* Paris: Recherches et Publications.

Kant, I. 1992. Concerning the Ultimate Ground of the Differentiation of Directions in Space. In *The Cambridge Edition of the Works of Kant: Theoretical Philosophy 1755–1770,* edited by D. Walford, 365–72. Cambridge, Engl.: Cambridge University Press.

Kapuściński, R. 1994. *Imperium.* New York: Knopf.

Kehoe, A. B. 1998. *The Land of Prehistory.* New York: Routledge.

Keith, K. E. 1999. *Cities, Neighborhoods and Houses: Urban Spatial Organization in Old Babylonian Mesopotamia.* Ph.D. dissertation, University of Michigan.

Kellner, H.-J. 1991. *Gürtelbleche aus Urartu.* Stuttgart: Franz Steiner Verlag.

Kemp, B. 1989. *Ancient Egypt: Anatomy of a Civilization.* London: Routledge.

Kennedy, P. M. 1987. *The Rise and Fall of the Great Powers: Economic Change and Military Conflict from 1500 to 2000.* 1st ed. New York: Random House.

Kenoyer, J. M. 1997. Early City-States in South Asia: Comparing the Harappan Phase and Early Historic Period. In *The Archaeology of City-States,* edited by D. L. Nichols and T. H. Charlton, 51–70. Washington, D.C.: Smithsonian Institution Press.

Kenyon, K. 1957. *Digging Up Jericho: the Results of the Jericho Excavations 1952–56.* New York: Praeger.

Khanzadian, E. V., K. A. Mkrtchian, and E. S. Parsamian. 1973. *Metsamor.* Yerevan: Akademiya Nauk Armianskoe SSR.

Khazanov, A. M. 1978. Some Theoretical Problems of the Study of the Early State. In *The Early State,* edited by H. J. M. Claessen and P. Skalnik, 77–92. The Hague: Mouton Publishers.

Khodzhash, S. I., N. S. Trukhtanova, and K. L. Oganesian. 1979. *Erebuni.* Moscow: Isskustvo.

Khoury, P. S., and J. Kostiner, eds. 1990. *Tribes and State Formation in the Middle East.* Berkeley: University of California Press.

Kirch, P. V. 1984. *The Evolution of the Polynesian Chiefdoms.* Cambridge, Engl.: Cambridge University Press.

———. 1988. *Niuatoputapu: The Prehistory of a Polynesian Chiefdom.* Seattle: Burke Museum.

Kitto, H. D. F. 1951. *The Greeks.* Harmondsworth: Penguin Books.

Kleiss, W. 1974. Planaufnahmen Urartäischer Burgen. *Archaeologische Mitteilungen aus Iran* 7:79–106.

———. 1977. *Bastam/Rusa-I-Uru.Tur.* Berlin: Dietrich Reimer Verlag.

———. 1982. Darstellungen Urartäischer Architektur. *Archaeologische Mitteilungen aus Iran* 15:53–77.

Knapp, A. B., ed. 1992. *Archaeology, Annales, and Ethnohistory.* Cambridge, Engl.: Cambridge University Press.

Kohl, P. L. 1978. The Balance of Trade in Southwestern Asia in the Mid-Third Millennium B.C. *Current Anthropology* 19(3):463–76.

———. 1989. The Use and Abuse of World Systems Theory: The Case of the "Pristine" West Asian State. In *Archaeological Thought in America,* edited by C. C. Lamberg-Karlovsky, 218–40. Cambridge, Engl.: Cambridge University Press.

Kohn, H. 1962. *The Age of Nationalism: The First Era of Global History.* New York: Harper.

Kolata, A. L. 1997. Of Kings and Capitals: Principles of Authority and the Nature of Cities in the Native Andean State. In *The Archaeology of City-States,* edited by D. L. Nichols and T. H. Charlton, 245–54. Washington, D.C.: Smithsonian Institution Press.

König, F. W. 1955. *Handbuch der Chaldischen Inschriften.* Graz: Selbstverlage des Herausgebers.

Kopytoff, I. 1987. The Internal African Frontier: The Making of African Political Culture. In *The African Frontier,* edited by I. Kopytoff, 2–84. Bloomington: Indiana University Press.

Kostoff, S. 1991. *The City Shaped*. London: Thames and Hudson.

Kouchoukos, N. 1999. *Landscape and Social Change in Late Prehistoric Mesopotamia*. Ph.D. dissertation, Yale University.

Kowalski, J. K., and N. P. Dunning. 1999. The Architecture of Uxmal: The Symbolics of Statemaking at a Puuc Maya Regional Capital. In *Mesoamerican Architecture as a Cultural Symbol,* edited by J. K. Kowalski, 274–97. New York: Oxford University Press.

Krader, L. 1968. *Formation of the State*. Englewood Cliffs, N.J.: Prentice-Hall.

Kramer, C. 1982. *Village Ethnoarchaeology: Rural Iran in Archaeological Perspective*. New York: Academic Press.

Kramer, S. 1963. *The Sumerians: Their History, Culture, and Character*. Chicago: University of Chicago Press.

Kroll, S. 1984. Urartus Untergang in Anderer Sicht. *Istanbuler Mitteilungen* 34:151–70.

Kruglov, A. P., and G. V. Podgayetsky. 1935. *Rodovoe Obshchestvo Stepei Vostochnoi Evropy*. Leningrad: Izvestiia GAIMK.

Kuhrt, A. 1995. *The Ancient Near East, c. 3000–330 B.C.* London: Routledge.

Kuklick, B. 1996. *Puritans in Babylon: The Ancient Near East and American Intellectual Life, 1880–1930*. Princeton: Princeton University Press.

Kunstler, J. H. 1993. *The Geography of Nowhere*. New York: Touchstone.

Kuper, H. 1972. The Language of Sites in the Politics of Space. *American Anthropologist* 74:411–25.

Kus, S. 1989. Sensuous Human Activity and the State: Towards an Archaeology of Bread and Circuses. In *Domination and Resistance,* edited by D. Miller, M. Rowlands, and C. Tilley, 140–54. London: Routledge.

———. 1992. Toward an Archaeology of Body and Soul. In *Representations in Archaeology,* edited by J.-C. Gardin and C. S. Peebles, 168–77. Bloomington: Indiana University Press.

Kus, S., and V. Raharijaona. 2000. House to Palace, Village to State: Scaling Up Architecture and Ideology. *American Anthropologist* 102(1):98–113.

Lacy, B. N. 1978. Introduction. In *The Federal Presence: Architecture, Politics, and Symbols in United States Government Building,* edited by L. A. Craig, vii–ix. Cambridge, Mass.: MIT Press.

Lamberg-Karlovsky, C. C. 1989. Mesopotamia, Central Asia, and the Indus Valley: So the Kings Were Killed. In *Archaeological Thought in America,* edited by C. C. Lamberg-Karlovsky, 241–67. Cambridge, Engl.: Cambridge University Press.

Lampl, P. 1968. *Cities and Planning in the Ancient Near East*. New York: George Braziller.

Lanfranchi, G. B., and S. Parpola, eds. 1990. *The Correspondence of Sargon II, Part II: Letters from the Northern and Northeastern Provinces*. Helsinki: Helsinki University Press.

Larsen, M. T. 1996. *The Conquest of Assyria: Excavations in an Antique Land, 1840–1860*. London: Routledge.

Lattimore, O. 1940. *Inner Asian Frontiers of China*. New York: American Geographical Society.

Layard, A. H. 1970. *Nineveh and Its Remains*. London: Routledge and Kegan Paul.

Le Corbusier. 1970. *Towards a New Architecture*. New York: Praeger.

Lefebvre, H. 1976. Reflections on the Politics of Space. *Antipode* 8:30–37.

———. 1991. *The Production of Space*. Oxford: Basil Blackwell.

Lenin, V. I. 1965. *The State: A Lecture Delivered at the Sverdlov University, July 11, 1919*. Peking: Foreign Languages Press.

Lenski, G. E. 1966. *Power and Privilege: A Theory of Social Stratification*. New York: McGraw-Hill.

Leone, M. P. 1988. The Georgian Order as the Order of Merchant Capitalism in Annapolis, Maryland. In *The Recovery of Meaning*, edited by M. P. Leone and P. B. Potter, 235–62. Washington, D.C.: Smithsonian Institution Press.

Levine, L. D. 1977. Sargon's Eighth Campaign. In *Mountains and Lowlands: Essays in the Archaeology of Greater Mesopotamia*, edited by L. D. Levine and T. C. Young Jr., 135–51. Malibu, Calif.: Undena.

Lewis, M. W., and K. E. Wigen. 1997. *The Myth of Continents: A Critique of Metageography*. Berkeley: University of California Press.

Lichtheim, M. 1976. *Ancient Egyptian Literature, A Book of Readings II: The New Kingdom*. Berkeley: University of California Press.

Lincoln, B. 1994. *Authority: Construction and Corrosion*. Chicago: University of Chicago Press.

Lipset, S. M. 1959. Political Sociology. In *Sociology Today: Problems and Prospects*, edited by R. K. Merton, 81–114. New York: Basic Books.

Liverani, M. 1997. Ancient Near Eastern Cities and Modern Ideologies. In *Die Orientalische Stadt: Kontinuät, Wandel, Bruch*, edited by G. Wilhelm, 85–107. Saarbrucken: SDV Saarbrücker.

Livingstone, D. 1992. *The Geographical Tradition*. Oxford: Blackwell.

Lloyd, S. 1980. Architecture of Mesopotamia and the Ancient Near East. In *Ancient Architecture*, edited by S. Lloyd and H. W. Müller, 7–74. New York: Rizzoli International.

Lösch, A. 1954. *The Economics of Location*. New Haven: Yale University Press.

Low, S. M. 2000. *On the Plaza: The Politics of Public Space and Culture*. Austin: University of Texas Press.

Lowenthal, D. 1961. Geography, Experience and Imagination: Towards a Geographical Epistemology. *Annals of the Association of American Geographers* 51:241–60.

Lowie, R. H. 1927. *The Origin of the State*. New York: Harcourt Brace.

Lucero, L. J. 1999. Classic Lowland Maya Political Organization: A Review. *Journal of World Prehistory* 13(2):211–63.

Lukács, G. 1971. *History and Class Consciousness: Studies in Marxist Dialectics*. Cambridge, Mass.: MIT Press.

Lukes, S. 1974. *Power: A Radical View*. London: Macmillan.

Machiavelli, N. 1998. *The Prince*. Chicago: University of Chicago Press.

MacIver, R. M. 1926. *The Modern State*. London: Oxford University Press.

Mackinder, H. J. 1904. The Geographical Pivot of History. *Geographical Journal* 23: 421–37.

MacKinnon, C. A. 1989. *Toward a Feminist Theory of the State.* Cambridge, Mass.: Harvard University Press.

Maekawa, K. 1973–74. The Development of the E-MI in Lagash During Early Dynastic III. *Mesopotamia* 8–9:77–144.

Mann, M. 1986. *The Sources of Social Power, vol. 1: A History of Power from the Beginning to A.D. 1760.* Cambridge, Engl.: Cambridge University Press.

Mannikka, E. 1996. *Angkor Wat: Time, Space, and Kingship.* Honolulu: University of Hawaii Press.

Mansfield, H. C. J. 1983. On the Impersonality of the Modern State: A Comment on Machiavelli's Use of Stato. *The American Political Science Review* 77(4):849–57.

March, J. G., and J. P. Olsen. 1989. *Rediscovering Institutions: The Organizational Basis of Politics.* New York: Free Press.

———. 1995. *Democratic Governance.* New York: Free Press.

Marchand, S. L. 1996. *Down from Olympus: Archaeology and Philhellenism in Germany, 1750–1970.* Princeton: Princeton University Press.

Marcus, J. 1973. Territorial Organization of the Lowland Classic Maya. *Science* 180(4089):911–16.

———. 1983. On the Nature of the Mesoamerican City. In *Prehistoric Settlement Patterns,* edited by E. Vogt and R. Leventhal, 195–242. Albuquerque: University of New Mexico Press.

———. 1992. *Mesoamerican Writing Systems: Propaganda, Myth, and History in Four Ancient Civilizations.* Princeton: Princeton University Press.

———. 1993. Ancient Maya Political Organization. In *Lowland Maya Civilization in the Eighth Century A.D.,* edited by J. A. Sabloff and J. S. Henderson, 111–83. Washington, D.C.: Dumbarton Oaks Research Library and Collection.

———. 1998. The Peaks and Valleys of Ancient States: An Extension of the Dynamic Model. In *Archaic States,* edited by G. Feinman and J. Marcus, 59–94. Santa Fe: School of American Research Press.

Marcus, J., and G. Feinman 1998. Introduction. In *Archaic States,* edited by G. Feinman and J. Marcus, 3–13. Santa Fe: School of American Research Press.

Margueron, J. 1982. *Recherches Sur les Palais Mesopotamiens de l'Age du Bronze.* Paris: Librarie Orientaliste Paul Geuthner S.A.

Markus, T. A. 1993. *Buildings and Power.* New York: Routledge.

Marling, K. A., ed. 1997. *Designing Disney's Theme Parks: The Architecture of Reassurance.* Montréal: Canadian Centre for Architecture.

Marlowe, C. 1999. *Tamburlaine the Great.* Edited by J. S. Cunningham. Manchester, Engl.: Manchester University Press.

Martin, S. and N. Grube. 1995. Maya Superstates. *Archaeology* 48(6):41–46.

———. 2000. *Chronicle of the Maya Kings and Queens: Deciphering the Dynasties of the Ancient Maya.* London: Thames and Hudson.

Martirosian, A. A. 1974. *Argishtihinili.* Yerevan: Izdatel'stvo Armyanskoj SSR.

Marx, K. 1986. *Karl Marx: A Reader.* Cambridge, Engl.: Cambridge University Press.

Marx, K., and F. Engels. 1998. *The German Ideology.* Amherst, N.Y.: Prometheus Books.

Massey, D. 1973. Towards a Critique of Industrial Location Theory. *Antipode* 5:33–39.

Matheny, R. T. 1987. An Early Maya Metropolis Uncovered: El Mirador. *National Geographic* 172(3):317–38.

Mathews, P. 1991. Classic Maya Emblem Glyphs. In *Classic Maya Political History,* edited by T. P. Culbert, 19–29. Cambridge, Engl.: Cambridge University Press.

Maudslay, A. P. 1889–1902. *Biologia Centrali-Americana.* London: R. H. Porter.

Mayo, J. M. 1988. *War Memorials as Political Landscape: The American Experience and Beyond.* New York: Praeger.

McGuire, R. H. 1983. Breaking Down Cultural Complexity: Inequality and Heterogeneity. *Advances in Archaeological Method and Theory* 6:91–142.

——. 1996. Why Complexity is Too Simple. In *Debating Complexity,* edited by D. A. Meyer, P. C. Dawson, and D. T. Hanna, 23–29. Calgary: The Archaeological Association of the University of Calgary.

McLeod, M. 1998. Architecture and Politics in the Reagan Era: From Postmodernism to Deconstructivism. In *Architecture Theory Since 1968,* edited by K. M. Hays, 680–702. Cambridge, Mass.: MIT Press.

Meinig, D. W., and J. B. Jackson. 1979. *The Interpretation of Ordinary Landscapes: Geographical Essays.* New York: Oxford University Press.

Melikishvili, G. A. 1951. Nekotorye Voprosy Social'no-ekonomiceskoj Istorii Nairi-Urartu. *Vestnik Drevnej Istorii* 4:22–40.

——. 1960. *Urartskie Klinoobraznye Nadpisi.* Moscow: Izdatel'stvo Akademii Nauk SSSR.

——. 1971. Urartskie Klinoobraznye Nadpisi, II. *Vestnik Drevnej Istorii* 4:267–94.

Mellart, J. 1967. *Çatal Hüyük. A Neolithic Town in Anatolia.* London: Thames and Hudson.

Michalowski, P. 1983. History as Charter: Some Observations on the Sumerian King List. *Journal of the American Oriental Society* 103(1):237–48.

——. 1987. Charisma and Control: On Continuity and Change in Early Mesopotamian Bureaucratic Systems. In *The Organization of Power: Aspects of Bureaucracy in the Ancient Near East,* edited by M. Gibson and R. Biggs, 55–68. Chicago: Oriental Institute.

——. 1989. *The Lamentation Over the Destruction of Sumer and Ur.* Winona Lake, Iowa: Eisenbrauns.

——. 1990. Early Mesopotamian Communicative Systems: Art, Literature, and Writing. In *Investigating Artistic Environments in the Ancient Near East,* edited by A. C. Gunter, 53–69. Washington, D.C.: Sackler Gallery Smithsonian Institution.

——. 1994. Writing and Literacy in Early States: A Mesopotamianist Perspective. In *Literacy: Interdisciplinary Conversations,* edited by D. Keller-Cohen, 49–70. Cresskill, N.J.: Hampton Press.

Mills, C. W. 1956. *The Power Elite.* New York: Oxford University Press.

Milton, J. 1971. History of Britain. In *Complete Prose Works, Volume V,* edited by F. Fogle. New Haven: Yale University Press.

Mitchell, M. 1998. *A New Kind of Party Animal: How the Young Are Tearing Up the American Political Landscape*. New York: Simon and Schuster.

Mitchell, T. P. 1991. The Limits of the State: Beyond Statist Approaches and Their Critics. *The American Political Science Revie* 85(1):77–96.

Mitchell, W. J. T. 1994. Imperial Landscape. In *Landscape and Power*, edited by W. J. T. Mitchell, 5–34. Chicago: University of Chicago Press.

———. 2000. Holy Landscape: Israel, Palestine, and the American Wilderness. *Critical Inquiry* 26(2):193–223.

Mollenkopf, J. H. 1992. *A Phoenix in the Ashes: The Rise and Fall of the Koch Coalition in New York City Politics*. Princeton: Princeton University Press.

Moran, W. 1995. The Gilgamesh Epic: A Masterpiece from Ancient Mesopotamia. In *Civilizations of the Ancient Near East*, edited by J. Sasson, 2327–36. New York: Scribners.

Morgan, L. H. 1985. *Ancient Society*. Tucson: University of Arizona Press.

Morley, S. G. 1946. *The Ancient Maya*. Stanford: Stanford University Press.

Movsisyan, A. Y. 1998. *The Hieroglyphic Script of Van Kingdom* [in Armenian]. Yerevan, Armenia: Gitutyun.

Mumford, L. 1937. What Is a City? *Architectural Record* 82(5):59–62.

Munn, N. D. 1986. *The Fame of Gawa: A Symbolic Study of Value Transformation in a Massim (Papua New Guinea) Society*. Cambridge, Engl.: Cambridge University Press.

———. 1992. The Cultural Anthropology of Time. *Annual Review of Anthropology* 21:93–123.

Nelson, L.-E. 2000. Fantasia. *New York Review of Books* 47 (8):4–7.

Nerlich, G. 1994. *The Shape of Space*. Cambridge, Engl.: Cambridge University Press.

Niskanen, W. A. 1971. *Bureaucracy and Representative Government*. Chicago: Aldine Atherton.

Nissen, H. 1988. *The Early History of the Ancient Near East, 9000–2000 B.C.* Chicago: University of Chicago Press.

Nivola, P. S. 1999. *Laws of the Landscape: How Policies Shape Cities in Europe and America*. Washington, D.C.: Brookings Institution.

Oakeshott, M. 1975. *On Human Conduct*. London: Clarendon Press.

Oates, J. 1986. *Babylon*. London: Thames and Hudson.

Oded, B. 1979. *Mass Deportations and Deportees in the Neo-Assyrian Empire*. Wiesbaden: Reichert.

Offe, C. 1984. *Contradictions of the Welfare State*. London: Hutchinson.

Offe, C., and V. Ronge. 1997. Theses on the Theory of the State. In *Contemporary Political Philosophy: An Anthology*, edited by R. E. Goodin and P. Pettit, 60–65. Oxford: Blackwell.

Oganesian, K. L. 1955. *Karmir-Blur IV: Arkitektura Teishabaini*. Yerevan: Akademii Nauk Armianskoj SSR.

———. 1961. *Arin-Berd I: Architektura Erebuni*. Yerevan: Akademii Nauk Armianskoj SSR.

———. 1973. *Rospisi Erebuni*. Yerevan: Akademii Nauk Armianskoj SSR.

Ögün, B. 1982. Die Urartäischen Paläste und die Bestattungsbräuche der Urartäer.

In *Beiträge zum Bauen und Wohnen im Altertum von Archäologen, Vor- und Frühgeschichten*, edited by H. Prückner, 217–35. Mainz: Philip von Zabern.

Olmsted, F. L. 1971. *Civilizing American Cities: A Selection of Frederick Law Olmsted's Writings on City Landscapes.* Cambridge, Mass.: MIT Press.

Olsen, D. J. 1986. *The City as a Work of Art.* New Haven: Yale University Press.

Olson, C. 1973 [1948]. Notes for the Proposition: Man Is Prospective. *Boundary* 2(1–2):2–3.

Oppenheim, A. L. 1950. Babylonian and Assyrian Historical Texts. In *Ancient Near Eastern Texts Relating to the Old Testament*, edited by J. B. Pritchard, 265–317. Princeton: Princeton University Press.

———. 1969. Mesopotamia—Land of Many Cities. In *Middle Eastern Cities*, edited by I. M. Lapidus. Berkeley: University of California Press.

Özgüç, T. 1966. *Altintepe.* Ankara, Turkey: Türk Tarih Kurumu.

———. 1969. Urartu and Altintepe. *Archaeology* 22(4):256–63.

Parsons, T. 1951. *The Social System.* Glencoe, Ill.: Free Press.

Peet, R. 1998. *Modern Geographical Thought.* Oxford: Blackwell.

Peters, B. G. 1999. *Institutional Theory in Political Science: The New Institutionalism.* London: Pinter.

Pierson, C. 1996. *The Modern State.* London: Routledge.

Piotrovskii, B. B. 1955. *Karmir-Blur III: Resultat Reskopok 1951–1953.* Yerevan: Akademii Nauk Armianskoj SSR.

———. 1959. *Vanskoe Tsartsvo (Urartu).* Moscow: Vostochnoe Literaturi.

———. 1967. *Urartu: The Kingdom of Van and Its Art.* Translated by Peter S. Gelling. New York: Praeger.

———. 1969. *The Ancient Civilization of Urartu.* New York: Cowles.

Piotrovskii, B. B., and L. T. Gyuzalian. 1933. Kreposti Armenii Dourartskogo i Urartskogo Vremeni. *Pamyatniki Istorii Materialnoi Kulturi* 5–6:57–60.

Place, V. 1867. *Ninive et l'Assyrie.* Paris: Imprimierie impériale.

Plato. 1941. *The Republic.* Translated by B. Jowett. New York: The Modern Library.

Poggi, G. 1990. *The State: Its Nature, Development, and Prospects.* Cambridge, Engl.: Polity.

Polanyi, K., C. M. Arensberg, and H. W. Pearson. 1957. *Trade and Market in the Early Empires.* Glencoe, Ill.: Free Press.

Pollock, S. 1999. *Ancient Mesopotamia.* Cambridge, Engl.: Cambridge University Press.

———. 2001. The Uruk Period in Southern Mesopotamia. In *Uruk Mesopotamia and Its Neighbors: Cross-Cultural Interactions in the Era of State Formation*, edited by M. S. Rothman, 181–231. Santa Fe: School of American Research Press.

Polsby, N. W. 1960. How to Study Community Power: The Pluralist Alternative. *The Journal of Politics* 22(3):474–84.

Popper, K. 1957. *The Poverty of Historicism.* London: Routledge and Kegan Paul.

Possehl, G. 1998. Sociocultural Complexity Without the State. In *Archaic States*, edited by G. Feinman and J. Marcus, 261–91. Santa Fe: School of American Research Press.

Postgate, J. N. 1992. *Early Mesopotamia: Society and Economy at the Dawn of History.* London: Routledge.

Postgate, N., T. Wang, and T. Wilkinson. 1995. The Evidence for Early Writing: Utilitarian or Ceremonial? *Antiquity* 69(264):459–80.

Poulantzas, N. 1973. *Political Power and Social Classes.* London: NLB and Sheed and Ward.

Powell, A. 1995. *Journals, 1982–1986.* London: Heinemann.

Ptolemy, C. 1991. *The Geography.* Mineola, N.Y.: Dover.

Rapoport, A. 1982. *The Meaning of the Built Environment.* Beverly Hills, Calif.: Sage Publications.

Rathje, W. L. 1971. The Origin and Development of Lowland Maya Civilization. *American Antiquity* 36:275–85.

Redfield, R. 1953. *The Primitive World and Its Transformations.* Ithaca, N.Y.: Cornell University Press.

Redman, C. L. 1978. *The Rise of Civilization.* San Francisco: W. H. Freeman.

———. 1999. *Human Impact on Ancient Environments.* Tucson: University of Arizona Press.

Renan, E. 1996. What Is a Nation? In *Becoming National: A Reader,* edited by G. Eley and R. G. Suny, 42–55. New York: Oxford University Press.

Renfrew, A. C. 1975. Trade as Action at a Distance: Questions of Integration and Communication. In *Ancient Civilization and Trade,* edited by J. A. Sabloff and C. C. Lamberg-Karlovsky, 3–59. Albuquerque: University of New Mexico Press.

———. 1986. Introduction: Peer-Polity Interaction and Socio-Political Change. In *Peer-Polity Interaction and Socio-Political Change,* edited by C. Renfrew and J. F. Cherry, 1–18. Cambridge, Engl.: Cambridge University Press.

Renfrew, C., and J. F. Cherry, eds. 1986. *Peer-Polity Interaction and Socio-Political Change.* Cambridge, Engl.: Cambridge University Press.

Riggs, C. R. 1999. *The Architecture of Grasshopper Pueblo: Dynamics of Form, Function, and Use of Space in a Prehistoric Community.* Ph.D. dissertation, University of Arizona.

Ritter, C. 1874. *Comparative Geography.* Philadelphia: Lippincott.

Roaf, M. 1990. *Cultural Atlas of Mesopotamia and the Ancient Near East.* New York: Facts On File.

Rosaldo, R. 1989. *Culture and Truth.* Boston: Beacon Press.

Roth, M. T. 1995a. *Law Collections from Mesopotamia and Asia Minor.* Atlanta, Ga.: Scholars Press.

———. 1995b. Mesopotamian Legal Traditions and the Laws of Hammurabi. *Chicago-Kent Law Review* 71(1):13–39.

Rothman, M. S. 1994. Sealings as a Control Mechanism in Prehistory: Tepe Gawra XI, X and VIII. In *Chiefdoms and Early States in the Near East: The Organizational Dynamics of Complexity,* edited by G. Stein and M. Rothman, 103–20. Madison, Wisc.: Prehistory Press.

———, ed. 2001. *Uruk Mesopotamia and Its Neighbors: Cross-Cultural Interactions in the Era of State Formation.* Santa Fe: School of American Research Press.

Rowen, H. 1961. 'L'etat, c'est a moi.' Louis XIV and the State. *French Historical Studies* 2:83–98.

Rowlands, M. J. 1989. A Question of Complexity. In *Domination and Resistance,* edited by D. Miller, M. J. Rowlands, and C. Tilley, 29–40. London: Unwin Hyman.

Rubin, Z. 1995. The Reforms of Khusro Anushirwan. In *The Byzantine and Early Islamic Near East, 3: States, Resources, and Armies. Papers of the Third Workshop on Late Antiquity and Early Islam*, edited by A. Cameron, 227–96. Princeton: Darwin Press.

Russell, J. M. 1991. *Sennacherib's Palace Without Rival at Nineveh*. Chicago: University of Chicago Press.

Ryan, W. B. F., and W. C. Pitman. 1998. *Noah's Flood: The New Scientific Discoveries About the Event that Changed History*. New York: Simon and Schuster.

Rykwert, J. 2000. *Seduction of Place: The City in the Twenty-First Century*. New York: Pantheon Books.

Sack, R. D. 1986. *Human Territoriality: Its Theory and History*. Cambridge, Engl.: Cambridge University Press.

Sahlins, M. 1976. *Culture and Practical Reason*. Chicago: University of Chicago Press.

Sahlins, M. D., and E. R. Service, eds. 1960. *Evolution and Culture*. Ann Arbor: University of Michigan Press.

Said, E. W. 1993. *Culture and Imperialism*. New York: Knopf.

Saitta, D. J. 1994. Agency, Class, and Archaeological Interpretation. *Journal of Anthropological Archaeology* 13(5):201–27.

Salvini, M. 1967. Nairi e Ur(u)atri. *Incunabula Graeca* 16.

———. 1969. Nuove Iscrizioni Urartea Dagli Scavi Di Arin-Berd, Nell'Armenia Sovietica. *Studi Micenei ed Egeo-Anatolici* 9:7–24.

———. 1989. Le Pantheon de l'Urartu et le Fondement de l'Etat. *Studi Epigrafici e Linguistici sul Vicino Oriente Antico* 6:79–89.

———. 1994. The Historical Background of the Urartian Monument Meher Kapısı. In *Anatolian Iron Ages 3*, edited by A. Çiligiroğlu and D. H. French, 211–20. Ankara, Turkey: The British Institute of Archaeology at Ankara.

———. 1995. *Geschichte und Kultur der Urartäer*. Darmstadt: Wissenschaftliche Buchgesellschaft.

Sanders, W. T. 1956. The Central Mexican Symbiotic Region: A Study in Prehistoric Settlement Patterns. In *Prehistoric Settlement Patterns in the New World*, edited by G. R. Willey, 115–27. New York: Wenner-Gren Foundation.

———. 1989. Household, Lineage, and State at Eighth-Century Copan, Honduras. In *The House of the Bacabs, Copan*, edited by D. L. Webster, 89–105. Washington, D.C.: Dumbarton Oaks.

Sanders, W. T., J. R. Parsons, and R. S. Santley. 1979. *The Basin of Mexico: Ecological Processes in the Evolution of a Civilization*. New York: Academic Press.

Sanderson, S. K. 1990. *Social Evolutionism: A Critical History*. Oxford: Blackwell.

———. 1995. *Social Transformations*. Oxford: Blackwell.

Santley, R. S. 1980. Disembedded Capitals Reconsidered. *American Antiquity* 45(1):132–45.

Sassen, S. 1994. *Cities in a World Economy*. Thousand Oaks, Calif.: Pine Forge Press.

Saunders, P. 1989. Space, Urbanism, and the Created Environment. In *Social Theory of Modern Societies: Anthony Giddens and His Critics,* edited by D. Held and J. B. Thompson, 215–34. Cambridge, Engl.: Cambridge University Press.

Sayre, W. S., and H. Kaufman. 1960. *Governing New York City: Politics in the Metropolis.* New York: Russell Sage Foundation.

Scarre, C., and B. M. Fagan. 1997. *Ancient Civilizations.* New York: Longman.

Schaeffer, F. K. 1953. Exceptionalism in Geography: A Methodological Examination. *Annals of the Association of American Geographers* 43:226–49.

Schama, S. 1996. *Landscape and Memory.* New York: Vintage Books.

Schele, L., and D. Freidel. 1990. *A Forest of Kings.* New York: Quill William Morrow.

Schele, L., and P. Mathews. 1998. *The Code of Kings: The Language of Seven Sacred Maya Temples and Tombs.* New York: Scribner.

Schiffer, M. B. 1983. Toward the Identification of Formation Processes. *American Antiquity* 48(4):675–706.

Schmandt-Besserat, D. 1996. *How Writing Came About.* Austin: University of Texas Press.

Schneider, A. 1920. *Die Sumerische Tempelstadt. Die Anfänge der Kulturwirtschaft.* Essen: Baedeker.

Schulze, R. 1958. The Role of Economic Dominants in Community Power Structure. *American Sociological Review* 23(1):3–9.

Schulze, R., and L. Blumberg. 1957. The Determination of Local Power Elites. *American Journal of Sociology* 63(3):290–96.

Schwartz, G. M., and S. E. Falconer. 1994. *Archaeological Views from the Countryside: Village Communities in Early Complex Societies.* Washington, D.C.: Smithsonian Institution Press.

Scott, J. C. 1998. *Seeing Like a State: How Certain Schemes to Improve the Human Condition Have Failed.* New Haven: Yale University Press.

Scully, V. J. 1991. *Architecture: The Natural and the Man-Made.* New York: St. Martin's Press.

Seidl, U. 1993. Urartäische Bauskulpturen. In *Aspects of Art and Iconography: Anatolia and Its Neighbors, Studies in Honor of N. Özgüç,* edited by M. Mellink, E. Porada, and T. Özgüç, 557–64. Ankara, Turkey: Türk Tarih Kurumu Basımevi.

Service, E. R. 1962. *Primitive Social Organization.* New York: Random House.

———. 1975. *Origins of the State and Civilization.* New York: Norton.

———. 1978. Classical and Modern Theories of the Origins of Government. In *Origins of the State: The Anthropology of Political Evolution,* edited by R. Cohen and E. R. Service, 21–34. Philadelphia: Institute for the Study of Human Issues.

Shamgar-Handelman, L., and D. Handelman. 1986. Holiday Celebrations in Israeli Kindergartens: Relationships Between Representations of Collectivity and Family in the Nation-State. In *The Frailty of Authority,* edited by M. J. Aronoff, 71–103. New Brunswick, N.J.: Transaction Books.

Shennan, S. 1993. After Social Evolution: A New Archaeological Agenda? In *Archaeological Theory: Who Sets the Agenda?,* edited by N. Yoffee and A. Sherrat, 53–59. Cambridge, Engl.: Cambridge University Press.

Sherratt, A. 1997. *Economy and Society in Prehistoric Europe.* Edinburgh: Edinburgh University Press.

Siu, H. F. 1986. Collective Economy, Authority, and Political Power in Rural China. In *The Frailty of Authority,* edited by M. J. Aronoff, 9–50. New Brunswick, N.J.: Transaction Books.

Skinner, G. W. 1977. Cities and the Hierarchy of Local Systems. In *The City in Late Imperial China,* edited by W. G. Skinner. Stanford: Stanford University Press.

Skinner, Q. 1997. The State. In *Contemporary Political Philosophy: An Anthology,* edited by R. E. Goodin and P. Pettit, 3–26. Oxford: Blackwell.

Sklar, L. 1974. *Space, Time, and Spacetime.* Berkeley: University of California.

Smith, A. D. 1991. *National Identity.* Reno: University of Nevada Press.

———. 1996. The Origins of Nations. In *Becoming National,* edited by G. Eley and R. G. Suny, 106–30. Oxford: Oxford University Press.

Smith, A. T. 1996. *Imperial Archipelago: The Making of the Urartian Landscape in Southern Transcaucasia.* Ph.D. dissertation, University of Arizona.

———. 1999. The Making of an Urartian Landscape in Southern Transcaucasia: A Study of Political Architectonics. *American Journal of Archaeology* 103(1):45–71.

———. 2000. Rendering the Political Aesthetic: Ideology and Legitimacy in Urartian Representations of the Built Environment. *Journal of Anthropological Archaeology* 19:131–63.

———. 2001. The Limitations of Doxa: Agency and Subjectivity from an Archaeological Point of View. *Journal of Social Archaeology* 1(2):155–71.

———. 2002. Urartian Spectacle: Authority, Subjectivity, and Aesthetic Politics. Paper presented in the workshop on "Spectacle, Performance, and Power in Premodern Complex Society," organized by T. Inomata and L. Coben at the Annual Meeting of the Society for American Archaeology, Denver.

Smith, A. T., R. Badaljan, and P. Avetissian. 1999. The Crucible of Complexity. *Discovering Archaeology* 1(2):48–55.

Smith, A. T., and P. Kohl. 1994. Coercion and Consent in the Rise of the Urartian State. Paper presented at the American Anthropological Association Annual Meeting, Atlanta, Ga.

Smith, A. T., and T. T. Thompson. Forthcoming. Urartu and the Southern Caucasian Political Tradition. In *A View from the Highlands,* edited by A. Sagona.

Smith, N. 1989. Uneven Development and Location Theory: Towards a Synthesis. In *New Models in Geography: The Political Economy Perspective,* edited by R. Peet and N. Thrift, 142–63. London: Unwin Hyman.

Soja, E. W. 1985. The Spatiality of Social Life: Towards a Transformative Retheorization. In *Social Relations and Spatial Structures,* edited by D. Gregory and J. Urry, 90–127. New York: St. Martin's Press.

———. 1988. *Postmodern Geographies: The Reassertion of Space in Critical Social Theory.* New York: Verso.

Soltan, K. E. 1996. Introduction: Imagination, Political Competence, and Institutions. In *The Constitution of Good Societies,* edited by K. E. Soltan and S. L. Elkin, 1–18. University Park: Pennsylvania State University Press.

Solzhenitsyn, A. I. 1985. *The Gulag Archipelago*. New York: Harper and Row.

Southall, A. W. 1953. *Alur Society: A Study in Processes and Types of Domination*. Cambridge, Engl.: Cambridge University Press.

———. 1965. A Critique of the Typology of States and Political Systems. In *Political Systems and the Distribution of Power*, 113–40. New York: Praeger.

———. 1998. *The City in Time and Space*. Cambridge, Engl.: Cambridge University Press.

Spencer, A. J. 1993. *Early Egypt: The Rise of Civilisation in the Nile Valley*. London: British Museum Press.

Spengler, O. 1932. *The Decline of the West*. Translated by C. F. Atkinson. New York: Knopf.

Stanford, M. 1998. *An Introduction to the Philosophy of History*. Oxford: Blackwell.

Stein, G. J. 1998. Heterogeneity, Power, and Political Economy: Some Current Research Issues in the Archaeology of Old World Complex Societies. *Journal of Archaeological Research* 6(1):1–44.

———. 1999. *Rethinking World-Systems: Diasporas, Colonies, and Interaction in Uruk Mesopotamia*. Tucson: University of Arizona Press.

———. 2001. Indigenous Social Complexity at Hacınebi (Turkey) and the Organization of Uruk Colonial Contact. In *Uruk Mesopotamia and Its Neighbors: Cross-Cultural Interactions in the Era of State Formation*, edited by M. S. Rothman, 265–305. Santa Fe: School of American Research Press.

Steinkeller, P. 1987. The Administrative and Economic Organization of the Ur III State. In *The Organization of Power: Aspects of Bureaucracy in the Ancient Near East*, edited by M. Gibson and R. D. Biggs, 19–41. Chicago: Oriental Institute of the University of Chicago.

Stern, R. 1980. The Doubles of Post-Modern. *Harvard Architecture Review* 1:74–87.

Steward, J. 1972. *Theory of Culture Change*. Urbana: University of Illinois Press.

Stewart, J. Q. 1947. Suggested Principles of Social Physics. *Science* 106:179–80.

Stigler, G. J. 1952. *The Theory of Price*. New York: Macmillan.

Stoker, G. 1996. Regime Theory and Urban Politics. In *The City Reader*, edited by R. T. LeGates and F. Stout, 268–81. London: Routledge.

Stone, A. 1999. Architectural Innovation in the Temple of the Warriors at Chichen Itza. In *Mesoamerican Architecture as a Cultural Symbol*, edited by J. K. Kowalski, 298–319. New York: Oxford University Press.

Stone, C. N. 1989. *Regime Politics: Governing Atlanta, 1946–1988*. Lawrence: University Press of Kansas.

Stone, E. 1987. *Nippur Neighborhoods*. Chicago: Oriental Institute.

———. 1995. The Development of Cities in Ancient Mesopotamia. In *Civilizations of the Ancient Near East*, edited by J. Sasson, 235–48. New York: Scribners.

Stone, E., and P. Zimansky. 1994. The Tell Abu-Duwari Project, 1988–1990. *Journal of Field Archaeology* 21(4):437–55.

Strong, W. D. 1953. Historical Approach in Anthropology. In *Anthropology Today*, edited by A. L. Kroeber, 386–97. Chicago: University of Chicago Press.

Szalay, M. 2000. *New Deal Modernism: American Literature and the Invention of the Welfare State*. Durham, N.C.: Duke University Press.

Tadgell, C. 1998. *Imperial Form: From Achaemenid Iran to Augustan Rome.* New York: Whitney Library of Design.

Tainter, J. A. 1988. *The Collapse of Complex Societies.* Cambridge, Engl.: Cambridge University Press.

Tambiah, S. J. 1977. The Galactic Polity: The Structure of Traditional Kingdoms in Southeast Asia. *Annals of the New York Academy of Sciences* 293:69–97.

Tate, C. E. 1992. *Yaxchilan: The Design of a Maya Ceremonial City.* Austin: University of Texas Press.

Taylor, P. J. 1997. World-Systems Analysis and Regional Geography. In *Political Geography,* edited by J. Agnew, 17–25. London: Arnold.

Thompson, E. P. 1963. *The Making of the English Working Class.* London: Gollancz.

Thompson, J. B. 1989. The Theory of Structuration. In *Social Theory of Modern Societies: Anthony Giddens and His Critics,* edited by D. Held and J. B. Thompson, 56–76. Cambridge, Engl.: Cambridge University Press.

Thompson, J. E. S. 1954. *The Rise and Fall of Maya Civilization.* Norman: University of Oklahoma Press.

Thongchai Winichakul. 1994. *Siam Mapped: A History of the Geo-Body of a Nation.* Honolulu: University of Hawaii Press.

Thünen, J. H. von. 1966. *Isolated State: An English Edition of Der Isolierte Staat.* Oxford: Pergamon Press.

Thurnwald, R., and H. Thurnwald. 1935. *Black and White in East Africa, The Fabric of a New Civilization: A Study of Social Contact and Adaptation of Life in East Africa.* London: G. Routledge and Sons.

Tilley, C. 1994. *A Phenomenology of Landscape.* Oxford: Berg.

Tilly, C. 1975. Reflections on the History of European State Making. In *The Formation of National States in Western Europe,* edited by C. Tilly, 3–83. Princeton: Princeton University Press.

Tinney, S. 1998. Death and Burial in Early Mesopotamia: The View from the Texts. In *Treasures from the Royal Tombs of Ur,* edited by R. L. Zettler and L. Horne, 26–31. Philadelphia: University of Pennsylvania Museum of Archaeology and Anthropology.

Tinniswood, A. 1998. *Visions of Power: Ambition and Architecture from Ancient Rome to Modern Paris.* London: M. Beazley.

Titus, C. H. 1931. A Nomenclature in Political Science. *American Political Science Review* 25:45–60.

Toramanyan, T. 1942. *Materials for the History of Armenian Architecture* [in Armenian]. Yerevan, Armenia: Haykakan SSR Gitowtyownneri Akademiayi.

Trigger, B. 1980. *Gordon Childe: Revolutions in Archaeology.* New York: Columbia University Press.

———. 1989. *A History of Archaeological Thought.* Cambridge, Engl.: Cambridge University Press.

———. 1998. *Sociocultural Evolution: Calculation and Contingency.* Oxford: Blackwell.

Tschumi, B. 1987. *Cinégram Folie, le Parc de la Villette.* Princeton: Princeton Architectural Press.

Tuan, Y.-F. 1977. *Space and Place: The Perspective of Experience.* Minneapolis: University of Minnesota Press.

Unger, R. M. 1997. *Politics: The Central Texts.* New York: Verso.

Urry, J. 1991. Time and Space in Giddens' Social Theory. In *Giddens' Theory of Structuration,* edited by G. Bryant and D. Jary, 160–75. London: Routledge.

Vajman, A. A. 1978. Urartskaya Ieroglifika: Rasshirovka Znaka i Chtenie Otdel'nych Nadpisej. In *Kultura Vostoka,* edited by V. G. Lukonin, 100–105. Leningrad: Avrora.

Van Buren, M., and J. E. Richards. 2000. Introduction: Ideology, Wealth, and the Comparative Study of "Civilizations." In *Order, Legitimacy, and Wealth in Ancient States,* edited by J. E. Richards and M. Van Buren, 3–12. Cambridge, Engl.: Cambridge University Press.

Van de Mieroop, M. 1992. *Society and Enterprise in Old Babylonian Ur.* Berlin: Dietrich Reimer Verlag.

———. 1997. *The Ancient Mesopotamian City.* Oxford: Clarendon Press.

Van Dyke, R. M. 1999. Space Syntax Analysis at the Chacoan Outlier of Guadalupe. *American Antiquity* 64(3):461–73.

Van Loon, M. 1966. *Urartian Art.* Istanbul: Nederlands Historisch-Archaeologisch Instituut.

———. 1975. The Inscription of Ishpuini and Meinua at Qalatgah, Iran. *Journal of Near Eastern Studies* 34(3):201–7.

Veenhof, K. R. 1980. Kaniš, Kārum. *Reallexion der Assyriologie* 5:369–78.

Venturi, R. 1966. *Complexity and Contradiction in Architecture.* New York: Museum of Modern Art.

Vico, G. 1996. *The New Science of Giambattista Vico.* New York: Legal Classics Library.

Vita-Finzi, C. 1978. *Archaeological Sites in Their Setting.* London: Thames and Hudson.

Viterbo, G. da. 1901. Liber de Regimine Civitatum. In *Bibliotheca Iuridica Medii Aevi,* vol. 3, edited by C. Salvemini, 215–80. Bologna: Societa Azzoguidiana.

Vogt, E. Z. 1983. Ancient and Contemporary Maya Settlement Patterns: A New Look from the Chiapas Highlands. In *Prehistoric Settlement Patterns: Essays in Honor of Gordon R. Willey,* edited by E. Z. Vogt and R. M. Leventhal, 89–114. Albuquerque: University of New Mexico Press.

Waldrop, M. M. 1992. *Complexity: The Emerging Science at the Edge of Order and Chaos.* New York: Simon and Schuster.

Wallerstein, I. 1974. *The Modern World System I: Capitalist Agriculture and the Origins of the European World-Economy in the Sixteenth Century.* San Diego, Calif.: Academic Press.

———. 1995. Hold the Tiller Firm: On Method and the Unit of Analysis. In *Civilizations and World Systems,* edited by S. K. Sanderson, 239–47. Walnut Creek, Calif.: Altamira Press.

Warnke, M. 1995. *Political Landscape: The Art History of Nature.* Cambridge, Mass.: Harvard University Press.

Wartke, R.-B. 1993. *Urartu: Das Reich am Ararat.* Mainz: Verlag Philipp von Zabern.

Weber, A. 1957. *Theory of the Location of Industries.* Chicago: University of Chicago Press.

Weber, M. 1946. *From Max Weber: Essays in Sociology.* Edited by H. H. Gerth and C. W. Mills. New York: Oxford University Press.

———. 1947. *Theory of Social and Economic Organization*. New York: Macmillan.

———. 1968. *Max Weber on Charisma and Institution Building*. Chicago: University of Chicago Press.

Weissleder, W. 1978. Aristotle's Concept of Political Structure and the State. In *Origins of the State*, edited by R. Cohen and E. R. Service, 187–203. Philadelphia: Institute for the Study of Human Issues.

Wenke, R. J. 1997. City-States, Nation States, and Territorial States: The Problem of Egypt. In *The Archaeology of City-States*, edited by D. L. Nichols and T. H. Charlton, 27–50. Washington, D.C.: Smithsonian Institution Press.

Werlen, B. 1993. *Society, Action, and Space: An Alternative Human Geography*. London: Routledge.

Wheatley, P. 1971. *The Pivot of the Four Quarters*. Edinburgh: Edinburgh University Press.

White, H. V. 1973. *Metahistory*. Baltimore: Johns Hopkins University Press.

White, L. 1949. *The Science of Culture*. New York: Farrar, Strauss.

Wilhelm, G. 1986. Urartu als Region der Keilschrift-Kultur. In *Das Reich Urartu: Ein Altorientalischer Staat im 1. Jahrtausend v. Chr.*, edited by V. Haas, 95–116. Konstanz: Universitatsverlag Konstanz GMBH.

Willey, G. R. 1981. Maya Lowland Settlement Pattern: A Summary Review. In *Lowland Maya Settlement Patterns*, edited by W. Ashmore, 385–415. Albuquerque: University of New Mexico Press.

Willey, G. R., W. R. J. Bullard, J. B. Glass, and J. C. Gifford. 1965. *Prehistoric Maya Settlement Patterns in the Belize Valley*. Cambridge, Mass.: Peabody Museum of Archaeology and Ethnology.

Wilson, J. A. 1958. The Report of a Frontier Official. In *The Ancient Near East*, vol. 1, edited by J. B. Pritchard, 183–84. Princeton: Princeton University Press.

Winter, I. J. 1983. The Program of the Throneroom of Assurnasirpal II. In *Essays on Near Eastern Art and Archaeology in Honor of Charles Kyrle Wilkinson*, edited by P. O. Harper and H. Pittman, 15–31. New York: Metropolitan Museum of Art.

———. 1993. "Seat of Kingship"/"A Wonder to Behold": The Palace as Construct in the Ancient Near East. *Ars Orientalis* 23:27–55.

Wirth, L. 1938. Urbanism as a Way of Life. *The American Journal of Sociology* 44(1):1–24.

Wise, M. Z. 1998. *Capital Dilemma: Germany's Search for a New Architecture of Democracy*. New York: Princeton Architectural Press.

Wittfogel, K. 1957. *Oriental Despotism: A Comparative Study of Total Power*. New Haven: Yale University Press.

Wolf, E. 1982. *Europe and the People Without History*. Berkeley: University of California Press.

———. 1984. Culture: Panacea or Problem. *American Antiquity* 49:393–400.

Woolley, C. L. 1939. *Ur Excavations: The Ziggurat and Its Surroundings*. London: Trustees of the British Museum and the University Museum, University of Pennsylvania.

———. 1965. *The Sumerians*. New York: W. W. Norton.

———. 1974. *Ur Excavations: The Buildings of the Third Dynasty*. London: Trustees of the British Museum and the University Museum, University of Pennsylvania.

Woolley, C. L., and M. Mallowan. 1976. *Ur Excavations: The Old Babylonian Period*. London: British Museum.

Wright, H. T. 1977. Recent Research on the Origin of the State. *Annual Review of Anthropology* 6:379–97.

———. 1978. Toward an Explanation of the Origin of the State. In *Origins of the State: The Anthropology of Political Evolution*, edited by R. Cohen and E. R. Service, 49–68. Philadelphia: Institute for the Study of Human Issues.

———. 1994. Prestate Political Formations. In *Chiefdoms and Early States in the Near East: The Organizational Dynamics of Complexity*, edited by G. Stein and M. Rothman, 67–84. Madison, Wisc.: Prehistory Press.

Wright, H. T., and G. A. Johnson. 1975. Population, Exchange, and Early State Formation in Southwestern Iran. *American Anthropologist* 77:267–89.

Wright, H. T., and E. S. A. Rupley. 2001. Calibrated Radiocarbon Age Determinations of Uruk-Related Assemblages. In *Uruk Mesopotamia and Its Neighbors: Cross-Cultural Interactions in the Era of State Formation*, edited by M. S. Rothman, 85–122. Santa Fe: School of American Research Press.

Wright, Richard. 1948. Two Letters to Dorothy Norman. In *Art and Action*, edited by D. Norman, 65–73. New York: Twice a Year Press.

Wright, Robert. 2000. *Nonzero: The Logic of Human Destiny*. New York: Pantheon Books.

Wylie, A. 1999. Why Should Archaeologists Study Capitalism? The Logic of Question and Answer and the Challenge of Systemic Analysis. In *Historical Archaeologies of Capitalism*, edited by M. P. Leone and P. B. Potter. New York: Plenum Press.

Yamin, R., and K. B. Metheny, eds. 1996. *Landscape Archaeology*. Knoxville: University of Tennessee Press.

Yoffee, N. 1979. The Decline and Rise of Mesopotamian Civilization: An Ethnoarchaeological Perspective on the Evolution of Social Complexity. *American Antiquity* 44:1–35.

———. 1993. Too Many Chiefs? or Safe Texts for the 90's. In *Archaeological Theory— Who Sets the Agenda*, edited by A. Sherratt and N. Yoffee, 60–78. Cambridge, Engl.: Cambridge University Press.

———. 1995. Political Economy in Early Mesopotamian States. *Annual Review of Anthropology* 24:281–311.

———. 1997. The Obvious and the Chimerical: City-States in Archaeological Perspective. In *The Archaeology of City-States: Cross-Cultural Approaches*, edited by D. L. Nichols and T. H. Charlton, 255–63. Washington, D.C.: Smithsonian Institution Press.

———. 2000. Law Courts and the Mediation of Social Conflict in Ancient Mesopotamia. In *Order, Legitimacy, and Wealth in Ancient States*, edited by J. E. Richards and M. Van Buren, 46–63. Cambridge, Engl.: Cambridge University Press.

Zettler, R. L. 1998. Ur of the Chaldees. In *Treasures from the Royal Tombs of Ur,* edited by R. L. Zettler and L. Horne, 9–19. Philadelphia: University of Pennsylvania Museum of Archaeology and Anthropology.

Zimansky, P. E. 1985. *Ecology and Empire.* Chicago: Oriental Institute.

———. 1990. Urartian Geography and Sargon's Eighth Campaign. *Journal of Near Eastern Studies* 49(1):1–21.

———. 1995. An Urartian Ozymandias. *Expedition* 58(2):94–100.

———. 1998. *Ancient Ararat: A Handbook of Urartian Studies.* Delmar, N.Y.: Caravan Books.

Žižek, S. 1999. *The Ticklish Subject: The Absent Centre of Political Ontology.* New York: Verso.

Zukin, S. 1991. *Landscapes of Power: From Detroit to Disney World.* Berkeley: University of California Press.

Index

Text:	10/13 Galliard
Display:	Galliard
Indexer:	Barbara Roos
Compositor:	Integrated Composition Systems
Printer:	Sheridan Books, Inc.